EXPRESS REVIEW GUIDES

Algebra I

EXPRESS REVIEW GUIDES

Algebra I

LEARNINGEXPRESS ®

New York

Library of Congress Cataloging-in-Publication Data:

Express review guides. Algebra I.—1st ed.
　　p. cm.
　　ISBN: 978-1-57685-594-2
　1.　Algebra—Outlines, syllabi, etc.　I. LearningExpress (Organization)
II. Title: Algebra I.
　QA159.2.E969 2007
　512—dc22

2007008905

ISBN: 978-1-57685-594-2 (alk. paper)

Printed in the United States of America

9　8　7　6　5　4　3　2　1

First Edition

For more information or to place an order, contact LearningExpress at:
　　55 Broadway
　　8th Floor
　　New York, NY 10006

Or visit us at:
　　www.learnatest.com

Contents

EXPRESS REVIEW GUIDES

Algebra I

Math? Why, oh Why?

Here's a common scenario that teachers run into. In every math classroom, without a doubt, a student will ask, "But why do we need to learn math?" Or, better yet, "Will we *ever* use this stuff in real life?"

Well, the answer is yes—I swear! If you cook a big meal for your family, you rely on math to follow recipes. Decorating your bedroom? You'll need math for figuring out what size bookshelf will fit in your room. Oh, and that motorcycle you want to buy? You're going to need math to figure out how much you can pay every month toward the total payment.

People have been using math for thousands of years, across various countries and continents. Whether you're sailing a boat off the coast of the Dominican Republic or building an apartment in Moscow, you're using math to get things done. You might be asking, how can math be so universal? Well, if you think about it, human beings didn't invent math concepts— we discovered them.

And then, there's algebra. Learning algebra is a little like learning another language. It's not exactly like English or Spanish or Japanese. Algebra is a simple language, used to create mathematical models of real-world situations and to help you with problems you aren't able to solve using plain old arithmetic. Rather than using words, algebra relies on numbers, symbols, and variables to make statements about things. And because algebra uses the same symbols as arithmetic for adding, subtracting, multiplying, and dividing, you're already familiar with the basic vocabulary!

People from different areas of the world and different backgrounds can "speak" algebra. If you are well versed in this language, you can use it to make important decisions and perform daily tasks with ease. Algebra can help you to shop intelligently on a budget, understand population growth, or even bet on the horse with the best chance of winning the race.

OKAY, BUT WHY *EXPRESS REVIEW GUIDES*?

If you're having trouble in math, this book can help you get on the right track. Even if you feel pretty confident about your math skills, you can use this book to help you review what you've already learned. Many study guides tell you how to improve your math—this book doesn't just *tell* you how to solve math problems, it *shows* you how. You'll find page after page of strategies that work, and you are never left stranded, wondering how to get the right answer to a problem. We'll show you all the steps to take so that you can successfully solve every single problem, and see the strategies at work.

Sometimes, math books assume you can follow explanations of difficult concepts even when they move quickly or leave out steps. That's not the case with this book. In each lesson, you'll find step-by-step strategies for tackling the different kinds of math problems. Then, you're given a chance to apply what you've learned by tackling practice problems. Answers to the practice problems are provided at the end of each section, so you can check your progress as you go along. This book is like your own personal math tutor!

THE GUTS OF THIS GUIDE

Okay, you've obviously cracked open the cover of this book if you're reading these words. But let's take a quick look at what is lurking in the other chapters. This book includes:

- ➡ a 50-question benchmark pretest to help you assess your knowledge of the concepts and skills in this guide
- ➡ brief, focused lessons covering essential algebra topics, skills, and applications
- ➡ specific tips and strategies to use as you study

➥ a 50-question posttest followed by complete answer explanations to help you assess your progress

As you work through this book, you'll notice that the chapters are sprinkled with all kinds of helpful tips and icons. Look for these icons and the tips they provide. They include:

➥ *Fuel for Thought*: critical information and definitions that can help you learn more about a particular topic

➥ *Practice Lap*: quick practice exercises and activities to let you test your knowledge

➥ *Inside Track*: tips for reducing your study and practice time—without sacrificing accuracy

➥ *Caution!*: pitfalls to be on the lookout for

➥ *Pace Yourself*: try these extra activities for added practice

READY, SET, GO!

To best use this guide, you should start by taking the pretest. You'll test your math skills and see where you might need to focus your study.

Your performance on the pretest will tell you several important things. First, it will tell you how much you need to study. For example, if you got eight out of ten questions right (not counting lucky guesses!), you might need to brush up only on certain areas of knowledge. But if you got only five out of ten questions right, you will need a thorough review. Second, it can tell you what you know well (that is, which subjects you *don't* need to study). Third, you will determine which subjects you need to study in-depth and which subjects you simply need to refresh your knowledge.

REMEMBER. . .

THE PRETEST IS only practice. If you did not do as well as you anticipated, do not be alarmed and certainly do not despair. The purpose of the quiz is to help you focus your efforts so that you can *improve.* It is important to analyze your results carefully. Look beyond your score, and consider *why* you answered some questions incorrectly. Some questions to ask yourself when you review your wrong answers:

- Did you get the question wrong because the material was totally unfamiliar?
- Was the material familiar, but you were unable to come up with the right answer? In this case, when you read the right answer, it will often make perfect sense. You might even think, "I knew that!"
- Did you answer incorrectly because you read the question carelessly?

Next, look at the questions you got right and review how you came up with the right answer. Not all right answers are created equal.

- Did you simply know the right answer?
- Did you make an educated guess? An educated guess might indicate that you have some familiarity with the subject, but you probably need at least a quick review.
- Did you make a lucky guess? A lucky guess means that you don't know the material and you will need to learn it.

After the pretest, begin the lessons, study the example problems, and try the practice problems. Check your answers as you go along, so if you miss a question, you can study a little more before moving on to the next lesson.

After you've completed all the lessons in the book, try the posttest to see how much you've learned. You'll also be able to see any areas where you may need a little more practice. You can go back to the section that covers that skill for some more review and practice.

THE RIGHT TOOLS FOR THE JOB

BE SURE THAT you have all the supplies you need on hand before you sit down to study. To help make studying more pleasant, select supplies that you enjoy using. Here is a list of supplies that you will probably need:

- A notebook or legal pad
- Graph paper
- Pencils
- Pencil sharpener
- Highlighter
- Index or other note cards
- Paper clips or sticky note pads for marking pages
- Calendar or personal digital assistant (which you will use to keep track of your study plan)
- Calculator

As you probably realize, no book can possibly cover all of the skills and concepts you may be faced with. However, this book is not just about building an algebra base, but also about building those essential skills that can help you solve unfamiliar and challenging questions. The algebra topics and skills in this book have been carefully selected to represent a cross section of basic skills that can be applied in a more complex setting, as needed.

Pretest

By taking this pretest, you will get an idea of how much you already know and how much you need to learn about algebra.

This pretest consists of 50 questions and should take about one hour to complete. You should not use a calculator when taking this pretest; however, you may use scratch paper for your calculations.

When you complete this pretest, check your responses against the Answers section. If your answers differ from the correct answers, use the explanations that have been given to retrace your calculations. Along with each answer is a number that tells you which chapter of this book teaches you the math skill needed for that question.

1. Suppose a number n is multiplied by 3, and 5 is added to the result, and then the result of that has been divided by 2, to arrive at $\frac{(3n + 5)}{2}$. What operations, in the correct order, must be performed on $\frac{(3n + 5)}{2}$ to arrive back at n?

2. How many pennies are q quarters worth?

3. How many quarters are p pennies worth?

4. Evaluate $5 + 6 \cdot 2$.

5. Evaluate $12 - 3 + 4$.

6. Evaluate $21 - (6 - 2) \div 2$.

7. Is the sentence $2 + 2 = 5$ open or closed? Is it true or false?

8. What is the solution set of $x^2 = 100$?

9. What is the solution set of $x^2 < 100$?

10. What is the solution set of $3x + 4x = 7x + 1$?

11. Evaluate the expression $12b + 3a$ when $a = -3$ and $b = 2$.

12. $5ab^4 - ab^4 =$

13. If Raven travels at 65 miles per hour (mph) for 3.5 hours, how far does she travel? Use the formula $D = rt$.

14. Carl works at a coffee shop and receives a 15% employee discount on all regularly priced food and merchandise. He has to pay full price for any sale item that he purchases. Write a formula that represents the total amount, t, he will spend if he buys three sale items priced at s and two regularly priced merchandise priced at r.

15. Given the equation:
$0.5x + 2 = -13 + \frac{3}{4}x$
Solve for x.

16. Given the equation:
$-6(x + 6) = -3(4x + 3)$
Solve for x.

17. Given the equation:

$-2x + 4 - 13x = 7x - 2 + 14$

Solve for x.

18. An institution pays 80% of the listed price on all tech equipment they purchase. If the institution spent \$7,600 on tech equipment in 2006, what would the nondiscounted price have been?

19. Find the slope (m) and y-intercept (b): $y - \frac{1}{2}x = 3\frac{1}{2}$.

20. A dive resort rents scuba equipment at a weekly rate of \$150 per week and charges \$8 per tank of compressed air during the week of diving. If y represents the total cost for one week and x equals the number of tanks used during the week, write an equation to represent a diver's cost for one week of diving at the resort.

21. Use the elimination method to solve the following system of equations:
$2x + 7y = 45$
$3x + 4y = 22$

22. Divide, where $x \neq 2$:
$\frac{3x - 6}{x - 2}$

23. Divide, where $6a + 4b \neq 0$:
$\frac{36a^2 - 16b^2}{6a + 4b} =$

24. Divide, where $x \neq 11$:
$\frac{x^2 - x - 110}{x - 11} =$

25. Solve for x: $x^2 + 4x = 12$

26. Simplify $\frac{24x^3}{(2x)^2} + \frac{3x^5}{x^4} - \frac{(3ax)^2}{a^2x}$.

27. Simplify $\sqrt{\frac{3x^2}{4}}$.

28. Simplify $\frac{\sqrt{12xy}}{\sqrt{x}}$.

29. What is the slope and the y-intercept of the graph of the line with equation $\frac{1}{2}x + y = 16$?

30. Solve the following system of equations for x and y:

$x + 2y = 18$

$3x - 6 = 2y$

31. Solve the following system of equations for x and y:

$2x + 5y = 60$

$x = 3.5y$

32. Solve the systems of inequalities to determine which of the following points lies within the solution set.

$y \geq 2(x - 5)^2$

$y < 2$

    a. $(0,0)$

    b. $(5,2)$

    c. $(5,1)$

    d. $(1,5)$

    e. $(-5,1)$

33. What expression is equivalent to "ten more than three times a number"?

34. Three times the sum of a number and 5 is equal to 27. What is the number?

35. One integer is four times another. The sum of the integers is 5. What is the value of the lesser integer?

36. The sum of 3 times the greater integer and 5 times the lesser integer is 9. Three less than the greater equals the lesser. What is the value of the lesser integer?

37. The perimeter of a rectangle is 104 inches. The width is 6 inches less than 3 times the length. Find the width of the rectangle.

38. What is the solution of the inequality $\frac{2}{5}x \leq 18$?

39. Which of the following is the solution of the inequality $-6a - 4 \geq 8$?

    a. $a \geq -2$

    b. $a \leq -2$

    c. $a \geq 2$

    d. $a \leq 2$

    e. $a < -2$

40. For which of the inequalities is the point $(3, -2)$ a solution?

    a. $2y - x \geq 1$

    b. $x + y > 5$

    c. $3y < -3x$

    d. $9x - 1 > y$

41. Factor the polynomial $6x + 12$ completely.

42. Factor the polynomial $4x^2 + 4x$ completely.

43. Factor the polynomial $10b^3 + 5b - 15$ completely.

44. Factor the polynomial $c^2 - 16$ completely.

45. Factor the polynomial $16x^2 - 25$ completely.

46. Factor the polynomial $x^2 + 8x + 7$ completely.

47. Factor the polynomial $x^2 + x - 20$ completely.

48. Factor the polynomial $2x^2 - 5x - 3$ completely.

49. Solve for x: $x^2 + 2x - 24 = 0$

50. Solve for x: $2x^2 + 3x = 4$

ANSWERS

1. Given $\frac{(3n + 5)}{2}$, first multiply it by 2 and get $3n + 5$. Then, add -5 and get $3n$. Then, multiply by $\frac{1}{3}$, resulting in n. The three operations are the inverse of the three operations that were originally performed on n, and they are performed in the opposite order. For more help with this concept, see Chapter 3.

2. One quarter is worth 25 pennies and q quarters are worth $25q$ pennies. For more help with this concept, see Chapter 3.

3. One penny is worth $\frac{1}{25}$ quarter and p pennies are worth $\frac{p}{25}$ quarters. For more help with this concept, see Chapter 3.

4. Perform multiplication first: $5 + 6 \cdot 2 = 5 + 12$. Add to find the solution: 17. For more help with this concept, see Chapter 3.

5. Because there are no other operations, add and subtract in order from left to right: $12 - 3 + 4 = 9 + 4 = 13$. For more help with this concept, see Chapter 3.

6. Evaluate the parentheses first: $21 - (6 - 2) \div 2 = 21 - (4) \div 2$. Divide to get $21 - 2$. Subtract to get a final answer of 19. For more help with this concept, see Chapter 3.

7. The sentence $2 + 2 = 5$ is closed, because it has no variables. It is false. For more help with this concept, see Chapter 4.

8. The solution set of $x^2 = 100$ is $\{10, -10\}$. For more help with this concept, see Chapter 4.

9. The solution set of $x^2 < 100$ is $\{x \mid -10 < x < 10\}$. For more help with this concept, see Chapter 4.

10. $3x + 4x = 7x + 1$

    $7x = 7x + 1$ Collect terms.

    $0 = 1$ Add $-7x$ to both sides.

    The solution set is $\{\}$, or the empty set. For more help with this concept, see Chapter 4.

11. Substitute $a = -3$ and $b = 2$ into $12b + 3a$ to get:

    $= 12(2) + 3(-3)$

    $= 24 - 9$

    $= 15$

    For more help with this concept, see Chapter 3.

12. Subtract like terms: $5ab^4 - ab^4 = 4ab^4$. For more help with this concept, see Chapter 5.

13. Use $D = rt$ with D as the unknown, $t = 3.5$ hours, and $r = 65$ miles per hour.

$D = rt$

$D = (65)(3.5)$

$D = 227.5$ miles

For more help with this concept, see Chapter 10.

14. Sale items are regularly priced, so $3 \cdot s$ will represent their total price. The regularly priced merchandise will be 85% of the regular price (with a 15% discount, Carl will pay 85%), so their price will be $2(0.85r)$. The total would then equal the sum of all: $t = 3s + 2(0.85r)$. For more help with this concept, see Chapter 10.

15. Convert the fraction to a decimal: $0.5x + 2 = -13 + \frac{3}{4}x$ becomes $0.5x + 2 = -13 + 0.75x$. Combine like terms by first subtracting $0.5x$ from both sides: $0.5x + 2 - 0.5x = -13 + 0.75x - 0.5x$. You now have $2 = -13 + 0.25x$. Add 13 to both sides: $2 + 13 = -13 + 0.25x + 13$. You are left with $15 = 0.25x$. Divide both sides by 0.25: $\frac{15}{0.25} = \frac{0.25x}{0.25}$. So, the correct answer is $60 = x$. For more help with this concept, see Chapter 3.

16. Apply the distributive property (to both sides): $-6(x + 6) = -3(4x + 3)$ becomes $-6x - 36 = -12x - 9$. Add $12x$ to both sides: $-6x - 36 + 12x = -12x - 9 + 12x$. You now have $6x - 36 = -9$. Add 36 to both sides: $6x - 36 + 36 = -9 + 36$. You are left with $6x = 27$. Divide both sides by 6: $x = \frac{27}{6}$, which simplifies to $\frac{9}{2}$. For more help with this concept, see Chapter 3.

17. First, combine like terms on both sides: $-2x + 4 - 13x = 7x - 2 + 14$ becomes $-15x + 4 = 7x + 12$. Add $15x$ to both sides: $-15x + 4 + 15x = 7x + 12 + 15x$. You now have $4 = 22x + 12$. Subtract 12 from both sides: $4 - 12 = 22x + 12 - 12$. You are left with $-8 = 22x$. Divide both sides by 22: $\frac{x}{22} = \frac{-8}{22}$. The correct answer is $x = \frac{-4}{11}$. For more help with this concept, see Chapter 3.

18. To solve this question, ask yourself, 80% of what number equals $7,600? Replace the phrase *what number* with your variable, x:

80% of $x = 7,600$

$0.80x = 7,600$

$x = \frac{7,600}{0.80} = \$9,500$

For more help with this concept, see Chapter 10.

19. Put the equation in the proper form. Add $\frac{1}{2}x$ to both sides of the equation: $y + \frac{1}{2}x - \frac{1}{2}x = \frac{1}{2}x + 3\frac{1}{2}$. Simplify the equation: $y = \frac{1}{2}x + 3\frac{1}{2}$. The equation is in the proper $\frac{\text{slope}}{y\text{-intercept}}$ form. The slope $(m) = \frac{1}{2}\left(\frac{\text{change in } y}{\text{change in } x}\right)$. The y-intercept $(b) = 3\frac{1}{2}$. For more help with this concept, see Chapter 6.

20. Let y equal the total cost for equipment. Let x equal the number of tanks used during the week. The problem tells us that the cost would be equal to the weekly charge for gear rental plus 8 times the number of tanks used. A formula that would represent this information would be $y = 8x + 150$. (The y-intercept would be at $(0,150)$ and the slope $= 8$, or $\frac{8}{1}$.) For more help with this concept, see Chapter 6.

21. Transform the first equation by multiplying by 3: $3(2x + 7y = 45)$. Use the distributive property: $3(2x) + 3(7y) = 3(45)$. Simplify terms: $6x + 21y = 135$. Now, transform the second equation by multiplying by -2: $-2(3x + 4y = 22)$. Use the distributive property: $-2(3x) - 2(4y) = -2(22)$. Simplify terms: $-6x - 8y = -44$. Add the transformed first equation to the second: $0 + 13y = 91$. Now divide both sides by 13: $y = 7$. Substitute the value of y into one of the equations in the system and solve for x: $3x + 4(7) = 22$. Simplify terms: $3x + 28 = 22$; $3x = -6$; $x = -2$. The solution for the system of equations is $(-2,7)$. For more help with this concept, see Chapter 7.

22. First, factor the top by using the GCF method: $\frac{3(x-2)}{x-2}$. Now you can cancel an "$x - 2$" from the top and bottom to get: $\frac{3(x-2)}{x-2} = 3$. For more help with this concept, see Chapter 9.

23. Use the difference of two squares factoring method on the numerator prior to division:

$\frac{36a^2 - 16b^2}{6a + 4b} =$

$\frac{(6a + 4b)(6a - 4b)}{6a + 4b} =$

$6a - 4b$

For more help with this concept, see Chapter 9.

24. Use the trinomial factoring method on the numerator prior to division: $\frac{x^2 - x - 110}{x - 11}$ becomes $\frac{(x + 10)(x - 11)}{x - 11}$. Divide: $\frac{(x + 10)(x - 11)}{x - 11} = x + 10$. For more help with this concept, see Chapter 9.

25. First, rearrange so the equation equals 0: $x^2 + 4x = 12$ becomes $x^2 + 4x - 12 = 0$. Next, factor: $(x + 6)(x - 2) = 0$. Set each factor equal to 0: $(x + 6) = 0$, so $x = -6$. Because $(x - 2) = 0$, $x = 2$. For more help with this concept, see Chapter 9.

26. Multiply the exponents of each factor inside the parentheses by the exponent outside the parentheses: $\frac{24x^3}{2^2x^2} + \frac{3x^5}{x^4} - \frac{3^2a^2x^2}{a^2x}$. Evaluate the numerical coefficients and divide out common numerical factors in the terms: $\frac{6x^3}{x^2} + \frac{3x^5}{x^4} - \frac{9a^2x^2}{a^2x}$. When similar factors, or bases, are being divided, subtract the exponent in the denominator from the exponent in the numerator: $6x^{3-2} + 3x^{5-4} - 9a^{2-2}x^{2-1}$. Simplify the operations in the exponents: $6x^1 + 3x^1 - 9a^0x^1$. Any term to the power of zero equals 1: $6x + 3x - 9(1)x = 0x = 0$. For more help with this concept, see Chapter 5.

27. Although it looks complex, you can still begin by factoring the terms in the radical sign: $\sqrt{\frac{3x^2}{4}} = \sqrt{\frac{3 \cdot x \cdot x}{2 \cdot 2}} = \sqrt{3 \cdot \frac{x \cdot x}{2 \cdot 2}}$. Factoring out the squares leaves $\sqrt{3 \cdot \left(\frac{x \cdot x}{2 \cdot 2}\right)} = \frac{x}{2}\sqrt{3}$. The result can be written a few different ways: $\frac{x}{2}\sqrt{3} = \frac{\sqrt{3}x}{2} = \frac{\sqrt{3}}{2}x = \frac{1}{2}x\sqrt{3}$. For more help with this concept, see Chapter 5.

28. First, rationalize the denominator in this expression. Then see if it can be simplified any further. Multiply the expression by 1 in a form suitable for this purpose: $\frac{\sqrt{12xy}}{\sqrt{x}} = \frac{\sqrt{12xy}}{\sqrt{x}} \cdot \frac{\sqrt{x}}{\sqrt{x}}$. Use the product property of radicals to combine the radicands in the numerator. In the denominator, a square root times itself is the radicand by itself: $\frac{\sqrt{12xy}}{\sqrt{x}} \cdot \frac{\sqrt{x}}{\sqrt{x}} = \frac{\sqrt{12xyz}}{x}$. The x in the numerator and the denominator divides out, leaving $2\sqrt{3y}$. For more help with this concept, see Chapter 5.

29. Rearrange the equation $\frac{1}{2}x + y = 16$ into the form $y = mx + b$: $y = -\frac{1}{2}x + 16$. Therefore, $m = -\frac{1}{2}$ and $b = 16$. The slope is $-\frac{1}{2}$, and the y-intercept is $(0,16)$. For more help with this concept, see Chapter 6.

30. Rearrange the second equation so it is in the same form as the first: $3x - 6 = 2y$ becomes $3x - 2y = 6$. Next, line up the two equations, and notice that adding them will cancel out the y terms:

$$x + 2y = 18$$
$$+\ 3x - 2y =\ 6$$
$$4x\ \ \ \ \ \ = 24$$

Divide both sides by 4: $\frac{4x}{4} = \frac{24}{4}$. So, $x = 6$. Next, substitute $x = 6$ into one of the equations. Let's use the second equation: $3x - 6 = 2y$ becomes $3(6) - 6 = 2y$ and then $18 - 6 = 2y$. After subtracting, you get $12 = 2y$. Divide both sides by 2: $\frac{12}{2} = \frac{2y}{2}$. The correct answer is $y = 6$. For more help with this concept, see Chapter 7.

31. Because $x = 3.5y$, you can substitute $3.5y$ for x in the first equation:

$2x + 5y = 60$

$2(3.5y) + 5y = 60$

$7y + 5y = 60$

$12y = 60$

$y = 5$

When $y = 12$, x will equal

$x = 3.5y$

$x = 3.5(5)$

$x = 17.5$

For more help with this concept, see Chapter 7.

32. Given $y \geq 2(x - 5)^2$ and $y < 2$, let's first plot $y \geq 2(x - 5)^2$, which is a parabola. Compared with the parabola $y = x^2$, this graph will be moved five units to the right. It will also double the value of the number that gets squared. Some (x,y) values are

(3,8)

(4,2)

(5,0)

(6,2)

(7,8)

Because the sign in the inequality is $\geq$, the curve will be solid and the shaded area will be above the curve. Next, consider $y < 2$. This will be a dotted line located at $y = 2$. Because $y < 2$, the area under the line will be shaded. Next, actually shade in the area that corresponds to both inequalities. The only point that lies within the shaded region is (5,1), choice **c**. For more help with this concept, see Chapter 8.

33. The key words *more than* tell you to use addition, and *three times a number* translates to $3x$. Therefore, *ten more than three times a number* can be rewritten with the expression $3x + 10$. For more help with this concept, see Chapter 10.

34. Let x equal a number. In the sentence, *three times the sum of a number and 5* translates into $3(x + 5)$. The second part of the sentence sets this equal to 27. The equation is $3(x + 5) = 27$. Use the distributive property on the left side of the equal sign: $3x + 15 = 27$. Subtract 15 from both sides of the equation: $3x + 15 - 15 = 27 - 15$. This simplifies to $3x = 12$. Divide both sides by 3: $\frac{3x}{3} = \frac{12}{3}$. Therefore, the correct answer is 4. For more help with this concept, see Chapter 10.

35. Let x equal the lesser integer and let y equal the greater integer. The first sentence in the question gives the equation $y = 4x$. The second sentence gives the equation $x + y = 5$. Substitute $y = 4x$ into the second equation: $x + 4x = 5$. Combine like terms on the left side of the equation: $5x = 5$. Divide both sides of the equation by 5: $\frac{5x}{5} = \frac{5}{5}$. This gives a solution of $x = 1$, which is the lesser integer. For more help with this concept, see Chapter 10.

36. Let x equal the lesser integer and let y equal the greater integer. The first sentence in the question gives the equation $3y + 5x = 9$. The second sentence gives the equation $y - 3 = x$. Substitute $y - 3$ for x in the first equation: $3y + 5(y - 3) = 9$. Use the distributive property on the left side of the equation: $3y + 5y - 15 = 9$. Combine like terms on the left side: $8y - 15 = 9$. Add 15 to both sides of the equation: $8y - 15 + 15 = 9 + 15$. Divide both sides of the equation by 8: $\frac{8y}{8} = \frac{24}{8}$. This gives a solution of $y = 3$. Therefore, the lesser integer, x, is 3 less than y, so $x = 0$. For more help with this concept, see Chapter 10.

37. Let l equal the length of the rectangle and let w equal the width of the rectangle. Because the width is 6 inches less than 3 times the length, one equation is $w = 3l - 6$. The formula for the perimeter of a rectangle is $2l + 2w = 104$. Substituting the first equation into the perimeter equation for w results in $2l + 2(3l - 6) = 104$. Use the distributive property on the left side of the equation: $2l + 6l - 12 = 104$. Combine like terms on the left side of the equation: $8l - 12 = 104$. Add 12 to both sides of the equation: $8l - 12 + 12 = 104 + 12$. Divide both sides of the equation by 8: $\frac{8l}{8} = \frac{116}{8}$. Therefore, the length is $l = 14.5$ inches and the width is $w = 3(14.5) - 6 = 37.5$ inches. For more help with this concept, see Chapter 10.

38. Multiply both sides of the inequality by $\frac{5}{2}$ and simplify; $\frac{5}{2}(\frac{2}{5})x \leq \frac{5}{2}(18)$; $\frac{10}{10}x \leq \frac{5}{2}(\frac{18}{1})$. The solution is $x \leq 45$. For more help with this concept, see Chapter 8.

39. Solve for a as you would in an equation. Add 4 to both sides of the inequality: $-6a - 4 + 4 \geq 8 + 4$. Simplify: $-6a \geq 12$. Divide both sides of the inequality by -6: $\frac{-6a}{-6} \geq \frac{12}{-6}$. Remember to switch the direction of the inequality symbol because you are dividing both sides by a negative number: $a \leq -2$ (choice **b**). For more help with this concept, see Chapter 8.

40. Substitute the point $(3,-2)$ into each answer choice to find the true inequality. Because the point is $(3,-2)$, the x-value is 3 and the y-value is -2. Twenty-six is *greater than* -2, so choice **d** is a solution. For more help with this concept, see Chapter 8.

41. Factor out the greatest common factor of 6 from each term. The binomial becomes $6(x + 2)$. For more help with this concept, see Chapter 9.

42. Factor out the greatest common factor of $4x$. The binomial becomes $4x(x + 1)$. For more help with this concept, see Chapter 9.

43. Factor out the greatest common factor of 5. The binomial becomes $5(2b^3 + b - 3)$. For more help with this concept, see Chapter 9.

44. This binomial is the difference between two perfect squares. The square root of c^2 is c, and the square root of 16 is 4. Therefore, the factors are $(c - 4)(c + 4)$. For more help with this concept, see Chapter 9.

45. This binomial is the difference between two perfect squares. The square root of $16x^2$ is $4x$, and the square root of 25 is 5. Therefore, the factors are $(4x - 5)(4x + 5)$. For more help with this concept, see Chapter 9.

46. Because each of the answer choices is a binomial, factor the trinomial into two binomials. To do this, you will be doing a method that resembles **FOIL**, performed backward: **F**irst terms of each binomial multiplied, **O**uter terms of each multiplied, **I**nner terms of each multiplied, and **L**ast term of each binomial multiplied. **F**irst results in x^2, so the first terms must be $(x \ )(x \ )$. **O**uter added to the **I**nner combines to $8x$, and the **L**ast is 7, so you need to find two numbers that add to $+8$ and multiply to $+7$. These two numbers would have to be $+1$ and $+7$. Therefore, the factors are $(x + 1)(x + 7)$. For more help with this concept, see Chapter 9.

47. Because each of the answer choices is a binomial, factor the trinomial into two binomials. To do this, you will be doing a method that resembles **FOIL**, performed backward: **F**irst terms of each binomial multiplied, **O**uter terms of each multiplied, **I**nner terms of each multiplied, and **L**ast term of each binomial multiplied. **F**irst results in x^2, so the first terms must be $(x\ )(x\ )$. **O**uter added to the **I**nner combines to $1x$, and the **L**ast is -20, so you need to find two numbers that add to $+1$ and multiply to -20. These two numbers would have to be -4 and $+5$. Therefore, the factors are $(x - 4)(x + 5)$. For more help with this concept, see Chapter 9.

48. Because each of the answer choices is a binomial, factor the trinomial into two binomials. To do this, you will be doing a method that resembles **FOIL**, performed backward: **F**irst terms of each binomial multiplied, **O**uter terms of each multiplied, **I**nner terms of each multiplied, and **L**ast term of each binomial multiplied. **F**irst results in $2x^2$, so the first terms must be $(2x\ )(x\ )$. The **L**ast is -3, so you need to find two numbers that multiply to -3. The **O**uter added to the **I**nner combines to $-5x$, but in this case, it is a little different because of the $2x$ in the first binomial. When you multiply the Outer values, you are not just multiplying by x but by $2x$. The only factors of -3 are -1 and 3 or 1 and -3, so try these values to see which placement of them works. The two numbers would then have to be 1 and -3, placed as follows: $(2x + 1)(x - 3)$. For more help with this concept, see Chapter 9.

49. In order to solve a quadratic equation, make sure that the equation is in standard form. This equation is already in the correct form: $x^2 + 2x - 24 = 0$. Factor the left side of the equation: $(x - 4)(x + 6) = 0$. Set each factor equal to zero and solve for x: $x - 4 = 0$ or $x + 6 = 0$. The solution to this equation is $x = 4$ or $x = -6$. For more help with this concept, see Chapter 9.

50. Put the equation in standard form: $2x^2 + 3x - 4 = 0$. Because this equation is not factorable, use the quadratic formula by identifying the values of a, b, and c, and then substituting them into the formula. For this particular equation, $a = 2$, $b = 3$, and $c = -4$. The quadratic equation is $x = \frac{-b \pm \sqrt{b^2 - 4ac}}{2a}$. Substitute the values of a, b, and c to get $x = \frac{-3 \pm \sqrt{3^2 - 4(2)(-4)}}{2(2)}$. This simplifies to $x = \frac{-3 \pm \sqrt{9 + 32}}{4}$, which becomes $x = \frac{-3 \pm \sqrt{41}}{4}$. So, $x = \frac{-3}{4} \pm \frac{\sqrt{41}}{4}$. The solution is $\{\frac{-3}{4} - \frac{\sqrt{41}}{4}, \frac{-3}{4} + \frac{\sqrt{41}}{4}\}$. For more help with this concept, see Chapter 9.

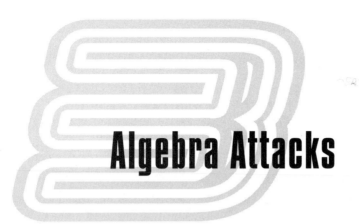

Algebra Attacks

Just when you were getting used to numbers, algebra attacks. You were hip to decimals, multiplication, even fractions. But algebra is a different kind of animal. It contains letters! Yes, letters. And that's not the only thing that makes algebra stand out.

You may be asking why you need to study algebra. Well, all the math that is needed in science and engineering relies on algebra as its basic language.

Do you like toasters and TVs? Do you like airplanes and air conditioning? Do you like all the smart, edgy stuff that technology makes available? It all depends on algebra.

LETTERS IN MATH CLASS?

Variables are letters that are used to represent numbers. Once you realize that these variables are just numbers in disguise, you'll understand that they must obey all the rules of mathematics, just like the numbers that aren't disguised. This can help you figure out what number the variable at hand stands for.

FUEL FOR THOUGHT

AN EXPRESSION IS like a series of words without a verb. Take, for example, $3x + 5$ or $a - 3$. A verb, which would be an equality or inequality symbol, would give value to the statement, turning it into an **equation** or **inequality**. These symbols include =, ≠, >, <, ≥, and ≤.

Look at this equation:

$a + 4 = 7$

In the equation $a + 4 = 7$, the letter a represents a little box. Think of the letter a as the label on the box. Inside the box is a number. Your job is to figure out what that number is. What number can you put in the box to turn the equation $a + 4 = 7$ into a true statement?

Let's just stick 20 into the a box. The equation becomes $20 + 4 = 7$, or $24 = 7$. Okay, this equation is *not* a true statement because 24 does not equal 7. So, a is not the number 20.

If you put 3 into the box, the equation becomes this true statement: $3 + 4 = 7$, or $7 = 7$.

FUEL FOR THOUGHT

WHEN A NUMBER is placed next to a variable, indicating multiplication, the number is said to be the **coefficient** of the variable. For example,

$8c$	8 is the coefficient of the variable c.
$6ab$	6 is the coefficient of both variables, a and b.

If two or more terms have exactly the same variable(s), they are said to be **like terms**.

$7x + 3x = 10x$

The process of grouping like terms together and performing their mathematical operations is called combining like terms. It is important to combine like terms carefully, making sure that the variables are exactly the same.

PRACTICE LAP

DIRECTIONS: Use scratch paper to solve the following problems. You can check your answers at the end of this chapter.

In questions 1 through 6, determine the expression, using the variable n for the number.

1. Three more than a number

2. Twice the sum of a number and three

3. Three more than twice a number

4. Three more than half a number

5. Half of three more than a number

6. Half of twice a number

7. Jenny was y years old three years ago. (a) How old is she now? (b) How old will she be five years from now?

8. Isabel can do a job in d days. What fraction of the job can she do in one day?

PRACTICE LAP

9. John and Mary are 20 miles apart. John is walking toward Mary at a speed of *j* miles per hour, and Mary is walking toward John at the rate of *m* miles per hour. How many miles closer together will John and Mary be in one hour?

10. How many pennies are *q* quarters worth?

11. What is the value in pennies of *n* nickels, *d* dimes, *q* quarters, and *b* dollar bills?

SIMPLIFYING EXPRESSIONS

In order to simplify expressions, there are several things you have to know.

To simplify $a - (b - c)$, just say no to the subtraction. Convert the expression to $a + -1(b - c)$, which then becomes $a + -b + c$ or $a - b + c$. Of course, you can go directly from $a - (b - c)$ to $a - b + c$ once you get the hang of it.

INSIDE TRACK

DON'T FORGET THE rules for order of operations with numerical expressions. You can use a memory device called a **mnemonic** to help you remember a set of instructions. Try remembering the acronym **PEMDAS**. This will help you remember to

P   do operations inside *Parentheses*

E   evaluate terms with *Exponents* (You will learn more about exponents in Chapter 5.)

M D   do *Multiplication* and *Division* in order from left to right

A S   *Add* and *Subtract* terms in order from left to right

When simplifying expressions, you should also start working as deeply inside the parentheses or braces as you can.

$3 - (4 - [5 - 2(a - b)]) =$
$3 - [4 - (5 - 2a + 2b)] =$
$3 - (4 - 5 + 2a - 2b) =$
$3 - [-1 + 2a - 2b] =$
$3 + 1 - 2a + 2b =$
$4 - 2a + 2b$

PRACTICE LAP

DIRECTIONS: Use scratch paper to solve the following problems. You can check your answers at the end of this chapter.

Simplify the following expressions.

12. $(3x + 4y - 7) - (5x - 3y - 7)$

13. $8x - (2 + 3[5 - (x - 2y)])$

SOLVING FOR THE VARIABLE

When you want to solve for a variable in an algebraic equation, you want to get your variable all by itself. We call this "isolating" the variable. In order to preserve the equality of the given equation, you need to be sure that you are doing the same thing to both sides of the equation. This means that you should perform corresponding operations on both sides of the equal sign. If you subtract 2 from the left side, you need to subtract 2 from the right side. If you divide the left side by 3, you must divide the right side by 3.

Example

Evaluate the expression $2b + a$ when $a = 2$ and $b = 4$.

Substitute 2 for the variable a and 4 for the variable b. When the expression is written as $2b$, it means 2 times b: $2(4) + 2$. Multiply $2 \cdot 4$. Finally, add $8 + 2 = 10$.

CAUTION!

WHEN TWO OR more numbers or variables are being multiplied, they are called **factors**. The answer that results is called the **product**.

$5 \times 6 = 30$ 5 and 6 are *factors*, and 30 is the *product*.

When you are working with variables, it may be easy to confuse the variable x with the multiplication sign $\times$. To avoid this, use the symbol $\cdot$ to represent the multiplication sign.

Multiplication is also indicated when a number is placed next to a variable. For example, look at $5a = 30$. In this equation, 5 is being multiplied by a.

Also, understand that parentheses around any part of one or more factors indicates multiplication: $(5)6 = 30$, $5(6) = 30$, and $(5)(6) = 30$.

PRACTICE LAP

DIRECTIONS: Use scratch paper to solve the following problems. You can check your answers at the end of this chapter.

14. Evaluate $a \div b - c - d$ when $a = -36$, $b = -9$, $c = 5$, and $d = -4$.

15. Solve the equation for x: $x - 3 = 12$.

16. Solve the equation for x: $\frac{x}{-4} = 11$.

17. Solve the equation for b: $3b - 11 = 52$.

18. Solve the equation for c: $15c - 12 - 3c = 36$.

19. Solve the equation for x: $8x - 24 = 6x$.

20. Solve the equation for p: $p - 3 = 4(3 - p)$.

21. Solve for a in terms of b and c: $11a - 6b = c$.

Algebra is governed by certain axioms. **Axioms** are statements that you regard as true because they are obvious and because your knowledge has to start somewhere. Other statements you need, beyond the axioms, can be proven based on the axioms.

THE BASIC AXIOMS

In this chapter, you will discover a basic set of axioms. Four axioms have to do with addition, and four axioms have to do with multiplication. The ninth axiom brings addition and multiplication together. The following chart offers a brief summary of these axioms.

	Addition	**Multiplication**
Commutative property	For any two numbers a and b, $a + b = b + a$.	For any two numbers a and b, $a \cdot b = b \cdot a$.
Associative property	For any three numbers a, b, and c, the expression $a + b + c$ can be interpreted either as $(a + b) + c$ or as $a + (b + c)$, and these two sums are equal.	For any three numbers a, b, and c, the expression $a \cdot b \cdot c$ can be interpreted either as $(a \cdot b) \cdot c$ or as $a \cdot (b \cdot c)$, and these two products are equal.
The identity element (a.k.a. the leave-it-alone number)	For any number a, $a + 0 = a$. Zero, added to any number, leaves that number alone. Zero is the identity element of addition. Zero preserves the identity of a, when you add zero to a.	For any number a, $a \cdot 1 = a$. One, multiplied by any number, leaves that number alone. One is the identity element of multiplication. One preserves the identity of a, when you multiply a by 1.
The existence of inverses	Every number (except zero) a has an additive opposite, $-a$, which satisfies $a + {-a} = 0$.	Almost every number a has a reciprocal $\frac{1}{a}$, which satisfies $a \cdot \frac{1}{a} = 1$. The exception is the number zero, which does not have a reciprocal.
Distributive property	For any three numbers a, p, and q, $a(p + q) = ap + aq$.	

THE COMMUTATIVE PROPERTY

The commutative property allows you to change the order of the numbers when you add or multiply.

Let's define addition, $a + b$, as repeated counting, so that $3 + 5$ means to start at 3 and count five times: 4, 5, 6, 7, 8. Then, $5 + 3$ means to start at 5 and count three times: 6, 7, 8. You can see that adding will come out to the same result both ways.

For multiplication, define $3 \cdot 5$ as repeated adding: 3 *times* 5 means to start with zero and *add* 5 three *times*: 0, 5, 10, 15. And 5 *times* 3 means to start with zero and *add* 3 five *times*: 0, 3, 6, 9, 12, 15. Remarkably, the result will always come out the same both ways.

And then, after seeing that the commutative property of addition and of multiplication makes sense for integers, you can make the leap and accept these laws for any numbers, whether they are integers or not.

THE ASSOCIATIVE PROPERTY

Adding is an operation you perform on two numbers at a time. So, the expression $3 + 4 + 5$ has to mean either:

$(3 + 4) + 5$ $(3 + 4,$ which is 7, added to 5)
or
$3 + (4 + 5)$ (the number 3 added to $4 + 5,$ which is 9)

In the first expression, the 4 "associates" with the 3; in the second expression, the 4 associates with the 5.

The associative property of addition guarantees that both expressions will yield the same number, regardless of which numbers you add first. In the first example, the result is $7 + 5 = 12$; in the second example, $3 + 9 = 12$. The same result, 12, occurs in both cases.

In the case of multiplication, $3 \cdot 4 \cdot 5$ can be interpreted either as

$(3 \cdot 4) \cdot 5$ $(3 \cdot 4,$ which is 12, multiplied by 5)
or
$3 \cdot (4 \cdot 5)$ (the number 3 multiplied by $4 \cdot 5,$ which is 20)

The associative property of multiplication asserts that the two examples will result in the same number: $12 \cdot 5 = 60$ and $3 \cdot 20 = 60$.

PRACTICE LAP

DIRECTIONS: Use scratch paper to solve the following problems. You can check your answers at the end of this chapter.

Show the two different interpretations, $(a + b) + c$ and $a + (b + c)$ for questions 22 and 23.

22. $5 + 6 + 7$

23. $20 + 30 + 40$

Show the two different interpretations, $(a \cdot b) \cdot c$ and $a \cdot (b \cdot c)$ for questions 24 and 25.

24. $5 \cdot 6 \cdot 7$

25. $20 \cdot 30 \cdot 40$

THE IDENTITY ELEMENT

Zero is the leave-it-alone number of addition—more formally known as the identity element of addition.

Adding zero to 17 results in 17: $17 + 0 = 17$. Expressed differently, adding zero to 17 preserves the identity of 17: $17 + 0 = 17$.

One is the leave-it-alone number of multiplication. Multiplying 17 by 1 leaves the 17 alone: $17 \cdot 1 = 17$. Expressed differently, multiplying 17 by 1 preserves the identity of 17: $17 \cdot 1 = 17$.

THE ADDITIVE AND THE MULTIPLICATIVE INVERSES

The additive inverse, or opposite, of a number n is the number that, when added to n, equals zero. The additive inverse of n is denoted $-n$. Every number has an additive inverse. The opposite of 3 is -3, and the opposite of -3 is 3. These numbers satisfy $3 + -3 = 0$.

If you take 17 and add 3 to get 20, you can get back to 17 by adding -3. And if you have 17 and you add -3 to get to 14, you can get back to 17 by adding 3. This is how the "oppositeness" of 3 and -3 works.

Multiplication works the same way. You get the multiplicative inverse by inverting the number. A number times its multiplicative inverse equals 1.

The reciprocal of 3 is $\frac{1}{3}$, and the reciprocal of $\frac{1}{3}$ is 3. Three and $\frac{1}{3}$ are reciprocals of each other. These numbers satisfy $3 \cdot \frac{1}{3} = 1$. Following are some other examples.

Examples

$$2 \cdot \frac{1}{2} = 1$$
$$\frac{3}{4} \cdot \frac{4}{3} = 1$$
$$-\frac{1}{5} \cdot -5 = 1$$

If you have 17 and you multiply by 3 to get 51, you can get back to 17 by multiplying 51 by $\frac{1}{3}$. And if you have 17 and multiply by $\frac{1}{3}$ to get $\frac{17}{3}$, you can get back to 17 by multiplying by 3. This is how the reciprocal nature of 3 and $\frac{1}{3}$ works.

These "do" and "undo" properties will aid you when you solve equations. Suppose you have the equation $3x + 4 = 19$.

Solving equations like this, which is a very important part of algebra, means to arrive at a statement that says, $x = \ldots$ (whatever it comes out to). In other words, the goal is to isolate the x.

On the left-hand side of this equation, x has been multiplied by 3, and 4 has been added. You can undo these operations by doing the inverse operations—adding -4 and multiplying by $\frac{1}{3}$.

$3x + 4 + -4 = 19 + -4$

$3x = 15$

$3x \cdot \frac{1}{3} = 15 \cdot \frac{1}{3}$

$x = 5$

That's how you can use inverses to your advantage in the world of algebra!

PRACTICE LAP

DIRECTIONS: Use scratch paper to solve the following problems. You can check your answers at the end of this chapter.

26. Suppose the way to Grandma's house from home is to walk north, turn left at the school, walk west, turn right at the church, and then walk north to Grandma's house. What is the way back? (Note that to get back home, you must do the inverse of each operation in the opposite order than you did them on the way to Grandma's house.)

27. Suppose you arrive at $\frac{(5x + 1)}{3}$ by multiplying x by 5, adding 1, and then multiplying by $\frac{1}{3}$. Can you get back to x by doing the three inverse operations in the opposite order? Explain why or why not.

THE DISTRIBUTIVE PROPERTY

What do you do with a problem like this: $2(x + y) + 3(x + 2y)$? According to the order of operations that you learned earlier in this chapter, you would have to do the symbols in parentheses first. However, you know you can't add x to y because they are not like terms. What you need to do is use the distributive property. The distributive property tells you to multiply the number and/or variable(s) outside the parentheses by every term inside the parentheses.

For this problem, multiply 2 by x and 2 by y. Then, multiply 3 by x and 3 by $2y$. If there is no number in front of the variable, it is understood to be 1, so 2 times x means 2 times $1x$. To multiply, you multiply the numbers and the variable stays the same. When you multiply 3 by $2y$, you multiply 3 by 2 and the variable, y, stays the same, so you would get $6y$. After you have multiplied, you can then combine like terms.

FUEL FOR THOUGHT

IF THERE IS no number in front of a variable, it is understood to be 1.

Example

$5(a + b) = 5a + 5b$

This can be proven by doing the math:

$5(1 + 2) = (5 \times 1) + (5 \times 2)$
$5(3) = 5 + 10$
$15 = 15$

PRACTICE LAP

DIRECTIONS: Use scratch paper to solve the following problem. You can check your answer at the end of this chapter.

28. Distribute r: $r(s + 2)$

ANSWERS

1. $n + 3$ is three more than n.

2. $n + 3$ is the sum of n and three, and $2(n + 3)$ is twice that.

3. $2n$ is twice n, and $2n + 3$ is three more than that.

4. Half of n is $\frac{n}{2}$, and $\frac{n}{2} + 3$ is three more than that.

5. $n + 3$ is three more than n, and $\frac{(n + 3)}{2}$ or $(\frac{1}{2})(n + 3)$ is half of that.

6. $2n$ is twice a number. You are looking for half of $2n$, or $(\frac{1}{2})2n$. Because $(\frac{1}{2}) \cdot 2$ equals 1, you can express half of twice a number as $1n$, or n.

7. (a) Jenny is $y + 3$ years old now. (b) Five years from now, she will be five years older than $y + 3$—that is, she will be $y + 3 + 5 = y + 8$ years old.

8. In one day, Isabel can do $\frac{1}{d}$ of the job. If this problem baffled you, the thing to do is to *de-algebra-ize* the problem. Suppose that Isabel can do a job in 4 days. What fraction of the job can she do in one day? You will realize the answer is $\frac{1}{4}$ of the job. You could also try supposing that Isabel can do the job in 5 days, and you will realize that she does $\frac{1}{5}$ of the job each day. Noting the trend, you will be able to make the leap back to algebra and realize that, in one day, Isabel can do $\frac{1}{d}$ of the job.

9. John and Mary will be $j + m$ miles closer together at the end of one hour. Although neither of them is traveling at the rate of $j + m$ miles per hour, the distance between them is diminishing at the rate of $j + m$ miles per hour. If the $j + m$ of this problem is eluding you, again *de-algebra-ize* the problem by having John walk at 3 miles per hour and Mary at 4 miles per hour. Draw a chart and plot out John's and Mary's progress during the hour. At the end of the hour, are they 7 miles closer to each other than they were originally?

10. q quarters are worth $25q$ pennies.

11. n nickels are worth $5n$ pennies; d dimes are worth $10d$ pennies; q quarters are worth $25q$ pennies; b dollar bills are worth $100b$ pennies. The whole stash is worth $5n + 10d + 25q + 100b$ pennies.

12. $(3x + 4y - 7) - (5x - 3y - 7) =$
    $3x + 4y + -7 + -5x + 3y + 7 =$
    $-2x + 7y$

13. $8x - (2 + 3[5 - (x - 2y)]) =$
    $8x - [2 + 3(5 + -x + 2y)] =$
    $8x - (2 + 15 + -3x + 6y) =$
    $8x - (17 + -3x + 6y) =$
    $8x + -17 + 3x + -6y =$
    $11x - 17 - 6y$

14. First, substitute numbers for the letters: $a = -36$, $b = -9$, $c = 5$, and $d = -4$ into $a \div b - c - d$: $-36 \div (-9) - 5 - (-4)$. Divide to get $(4) - 5 - (-4)$. Subtract from left to right: $4 - 5 = -1$, so the expression becomes $-1 - (-4)$. Change the subtraction to addition and the sign of the 4 to its opposite: $-1 + 4 = 3$.

15. Add 3 to both sides of the equation: $x - 3 + 3 = 12 + 3$. Simplify: $x = 15$.

16. Multiply each side of the equation by -4: $-4 \cdot \frac{x}{-4} = 11 \cdot -4$. Because the -4's on the left side cancel out, this leaves $x = -44$.

17. First, add 11 to both sides of the equation: $3b - 11 + 11 = 52 + 11$. This results in $3b = 63$. Divide both sides of the equation by 3: $\frac{3b}{3} = \frac{63}{3}$. The correct answer is $b = 21$.

18. Combine like terms on the left side of the equation: $12c - 12 = 36$. Add 12 to both sides of the equation: $12c - 12 + 12 = 36 + 12$. This simplifies to $12c = 48$. Divide both sides of the equation by 12: $\frac{12c}{12} = \frac{48}{12}$. The correct answer is $c = 4$.

19. Subtract $8x$ from both sides of the equation to get the variables on one side: $8x - 8x - 24 = 6x - 8x$. This simplifies to $-24 = -2x$. Divide both sides of the equation by -2: $\frac{-24}{-2} = \frac{-2x}{-2}$. The correct answer is $12 = x$.

20. Eliminate the parentheses: $p - 3 = 12 - 4p$. Add $4p$ to both sides of the equation: $p + 4p - 3 = 12 - 4p + 4p$. Combine like terms: $5p - 3 = 12$. Add 3 to both sides of the equation: $5p - 3 + 3 = 12 + 3$. You are left with $5p = 15$. Divide both sides by 5: $\frac{5p}{5} = \frac{15}{5}$. The correct answer is $p = 3$.

21. Add $6b$ to both sides of the equation: $11a - 6b + 6b = c + 6b$. Simplify: $11a = c + 6b$. Divide both sides of the equation by 11: $a = \frac{(c + 6b)}{11}$.

22. $(5 + 6) + 7 = 11 + 7 = 18; 5 + (6 + 7) = 5 + 13 = 18.$

23. $(20 + 30) + 40 = 50 + 40 = 90; 20 + (30 + 40) = 20 + 70 = 90.$

24. $(5 \cdot 6) \cdot 7 = 30 \cdot 7 = 210; 5 \cdot (6 \cdot 7) = 5 \cdot 42 = 210.$

25. $(20 \cdot 30) \cdot 40 = 600 \cdot 40 = 24{,}000; 20 \cdot (30 \cdot 40) = 20 \cdot 1{,}200 = 24{,}000.$

26. To get home, you need to do these five operations:

Walk south.

Turn left at the church.

Walk east.

Turn right at the school.

Walk south.

Notice that each of the five operations is the opposite of the operation you used to get to Grandma's house. Also, note that the inverse operations are done in the opposite order.

27. Start with $\frac{(5x + 1)}{3}$. Multiply by 3: $3(\frac{5x}{3} + \frac{1}{3}) = 5x + 1$. Add -1: $5x + 1 + -1 = 5x$. Multiply by $\frac{1}{5}$: $(5x)\frac{1}{5} = x$. You can get back to x by doing the three inverse operations in the opposite order.

28. You can distribute the r by multiplying it by each factor in the parentheses: $rs + 2r$.

Equations and Inequalities

WHAT'S AROUND THE BEND

➥ Closed Statements

➥ Open Statements

➥ Solution Sets

➥ Changing Expressions and Inequalities

As you read in Chapter 3, an expression has no equal sign and no sign of inequality ($\neq$, $<$, $\leq$, $>$, or $\geq$). An equation, on the other hand, consists of two expressions connected by an equal sign, like $3x + 5 = 20$ or $3 \cdot 10 = 30$. An inequality is two expressions connected by a sign of inequality, like $3x + 5 \neq 20$ or $3x > 12$.

An equation or inequality is said to be a **closed statement** if it has no variables. Take a peek at the following closed statements:

$3 > 2$
$2 + 2 = 4$

A closed statement is either true or false. The statement $3 > 2$ is an example of a closed statement that is true, because 3 is greater than 2. The statement $2 > 3$ is a closed statement that is false; 2 cannot be greater than 3.

An **open statement** is a statement that has one or more variables. An open statement, like $4x = 12$, is neither true nor false. Instead, an open statement has a solution set. The solution set for $4x = 12$ contains the number 3. The number 3, when substituted for x in $4x = 12$, transforms the open statement $4x = 12$ into a statement, $4 \cdot 3 = 12$, which is both closed and true. You would write the solution set for this open statement as {3}.

Let's look at open statements that have more than one number in their solution sets. The equation $x^2 = 9$ has two solutions, 3 and -3. The solution set is written {3,−3}. The solution set for the inequality $2x > 18$ is {all numbers greater than 9}.

Open sentences can have two or more variables. The open sentence $y = 3x + 4$ has two variables, x and y. The solution set consists not of numbers, but of number pairs. The number pair (0,4), which signifies $x = 0$ and $y = 4$, satisfies the equation, so it is a member of the solution set. Another member is (1,7). The solution set can be written as {(0,4), (1,7), . . .}. This solution set has an infinite number of members.

Following are several equations and their descriptions. Look carefully at each solution set.

	1	2	3	4	5	6	7
equation	$2+2=4$	$2+2=5$	$2+x=5$	$4x=12$	$x^2=9$	$y=3x+4$	$2x+3x=5x$
open or closed statement	closed	closed	open	open	open	open	open
number of variables	0	0	1	1	1	2	1
truth value	true	false	n/a	n/a	n/a	n/a	n/a
solution set	n/a	n/a	{3}	{3}	{3,−3}	{(0,4), (1,7),...}	{all real numbers}

Equation 1 is a closed statement. It has no variable, and it is true.

Equation 2 is also a closed statement, with no variable. It is also false: $2 + 2 = 4$, not 5.

The rest of the equations are open statements, because each one has one or more variables. As open statements, they lack a truth value. Rather, they have solution sets.

Equations 3 and 4 have only one solution, the number 3. The solution set for each of these two equations is {3}.

Equation 5 is an equation that has two solutions. Its solution set is {3,–3}.

Equation 6 has two variables. Remember, its solution set does not consist of numbers. The solution set consists of number pairs, including (0,4) and (1,7). You can verify this by substituting 1 for x and 7 for y. The solution set is {(0,4), (1,7), . . .}, which has an infinite number of members.

Equation 7 has one variable. This equation is true no matter what value is substituted for x. If $x = 10$, you have $2 \cdot 10 + 3 \cdot 10 = 5 \cdot 10$, or $20 + 30 = 50$. If $x = 11$, you have $2 \cdot 11 + 3 \cdot 11 = 5 \cdot 11$, or $22 + 33 = 55$. Both of these solutions are true. The solution set of this equation is all real numbers. An equation that is true for all real numbers is called an **identity**.

Following are several inequalities and their descriptions. Again, look carefully at each solution set.

	1	2	3	4	5
inequality	$3 > 2$	$2 > 3$	$2 + x > 5$	$x^2 < 9$	$y \geq 3x + 4$
open or closed statement	closed	closed	open	open	open
number of variables	0	0	1	1	2
truth value	true	false	n/a	n/a	n/a
solution set	n/a	n/a	$\{x \mid x > 3\}$	$\{x \mid -3 < x < 3\}$	$\{(0,4), (0,4.1), (1,7), (1,7.1), \ldots\}$

For expression 3, $\{x \mid x > 3\}$ means all real numbers x that satisfy $x > 3$.

- PRACTICE LAP -------------------

DIRECTIONS: Use scratch paper to solve the following problems. You can check your answers at the end of this chapter.

State whether the following statements are equalities or inequalities, whether they are open or closed, how many variables are in each statement, and if there is a solution set or not.

1. $2 + 2 = 5$
2. $5 < 4$
3. $2 + x = 5$
4. $3x + 4y = 5$
5. $x^2 + y^2 \leq 25$
6. $x^2 + y^2 = 25$
7. $2 + 2 = 4$

CHANGING EXPRESSIONS AND INEQUALITIES

In changing an equation, the objective is to avoid changing the solution set.

In the equation $3x^2 + 8 = 56$, the solution set is $\{4, -4\}$. When you add -8 to both sides, you change the values of both expressions (the left-hand side of the equation and the right-hand side) and get the new equation $3x^2 = 48$. The solution set of this equation is also $\{4, -4\}$. Likewise, when you change the value of both expressions by multiplying by $\frac{1}{3}$, you get the equation $x^2 = 16$, which also has the solution set $\{4, -4\}$.

You can add the same thing to both sides of an equation and, in so doing, be rest assured that the solution set did not change. You can multiply both sides of an equation by the same positive or negative number (but not zero) and, in so doing, be rest assured that the solution set does not change.

The object in changing inequalities, like in the case of equations, is to avoid changing the solution set. Whenever you add the same thing to both sides of an inequality, the solution set does not change.

For multiplication, the story is a little more involved. Three is less than 4 ($3 < 4$), meaning that 3 on the number line is to the left of 4. When you flip those numbers to their inverses so that you are looking at -3 and -4, you find that −4 is to the left of −3. The order changes when you multiply by −1 or by any negative number.

You can multiply both sides of an inequality by the same positive number and the solution set does not change. However, when you multiply both sides by the same negative number, you have to change the sense of the inequality to assure that the solution set does not change. By changing the sense of the inequality, you are making these changes:

< to >
> to <
≤ to ≥
≥ to ≤

CAUTION!

BEWARE—SOMETIMES YOU will change the solution set. Can you square both sides of an equation? Yes. After all, who is going to stop you? The question is, when you square both sides of an equation, does the solution set change? Well, it is a possibility.

Suppose you have $2x = 10$. The solution set is {5}. If you square both sides, getting $4x^2 = 100$, the solution set is {5,−5}. The solution set did change—a number that wasn't in the solution set of the original equation found its way into the solution set of the squared equation. A number that creeps into the solution set of an equation in that manner is called a **spurious solution** or an **extraneous solution**.

What does this all mean? Well, you *can* square both sides of an equation. In doing so, rest assured that you didn't lose any elements from the solution set. But, additional elements may have crept in.

ANSWERS

1. equality; closed; no variables; no solution set
2. inequality; closed; no variables; no solution set
3. equality; open; one variable; there is a solution set
4. equality; open; two variables; there is a solution set
5. inequality; open; two variables; there is a solution set
6. equality; open; two variables; there is a solution set
7. equality; closed; no variables; no solution set

Exploring Exponents and Radicals

WHAT'S AROUND THE BEND

- Exponents
- Bases
- Powers
- Adding and Subtracting
  with Exponents
- Multiplying with Exponents
- Dividing with Exponents
- Radicals
- Radical Signs
- Radicands
- Perfect Squares
- Simplifying Radicals
- Radicands and Fractions
- Rationalizing Denominators
- Adding and Subtracting Square Roots
- Multiplying and Dividing Radicals
- Solving Radical Equations

You may recall that addition is repeated counting. The expression 5 + 3 means start at 5 and count three times: 6, 7, 8.

Multiplication is repeated adding. The expression 5 · 3 means start at zero and add 3 five times: 0 + 3 + 3 + 3 + 3 + 3, or 0, 3, 6, 9, 12, 15.

If you use repeated multiplication, you are working with exponents. The expression 3^5 (said as "3 to the power of 5" or "3 to the fifth power") means start with 1 and multiply by 3 five times: 1 · 3 · 3 · 3 · 3 · 3, or 1, 3, 9, 27, 81, 243.

An exponent tells you how many times a factor is multiplied. An exponent appears as a raised number that is smaller in size than the other numbers. For example, in the expression 4^3, 3 is the exponent. The expression 4^3 shows that 4 is a factor three times. That means 4 · 4 · 4. Here are examples of exponents and what they mean:

$$5^2 = 5 \cdot 5$$
$$2^3 = 2 \cdot 2 \cdot 2$$
$$a^2 = a \cdot a$$
$$2x^3y^2 = 2 \cdot x \cdot x \cdot x \cdot y \cdot y$$

FUEL FOR THOUGHT

WHEN YOU WRITE b^m for a positive number b, b is the **base** and m is the **exponent**. You say that b^m is "b to the mth power" or "b to the power of m." For the **powers** 2 and 3, you say that b^2 is "b squared" and b^3 is "b cubed."

Again, when you write b^m for positive b, you have "raised b to the power of m." This process is called **exponentiation**.

In the order of operations, you should do exponents before moving on to multiplication, division, addition, or subtraction.

The expression $2 \cdot 3^4$ should be interpreted as 2 · (3 · 3 · 3 · 3) = 2 · 81 = 162.

If you want to square the quantity ab, writing ab^2 doesn't do it because the exponent in ab^2 squares only b. What ab^2 means is $a \cdot b \cdot b$. In order to square the quantity ab, you need to use parentheses: $(ab)^2$, which means $(ab)(ab) = a \cdot b \cdot a \cdot b = a \cdot a \cdot b \cdot b = a^2b^2$.

ADDING AND SUBTRACTING WITH EXPONENTS

When you combine similar terms, you add the numbers in front of the variables (coefficients) and leave the variables the same. Here are some examples:

$3x + 4x = 7x$
$2x^2 + 7x^2 = 9x^2$
$3xy + 6xy = 9xy$
$5x^3 - 3x^3 = 2x^3$

What do you do with exponents when you are adding or subtracting? Nothing! That's right—you add or subtract the coefficients only. The variables and their exponents stay the same.

MULTIPLYING WITH EXPONENTS

The rules for multiplying expressions with exponents may appear confusing. You treat exponents differently from ordinary numbers. You would think that when you are multiplying, you would multiply the exponents. However, that's not true. When you are multiplying expressions, you *add* the exponents. Here's an example of how to simplify an expression.

Example
$x^2 \cdot x^3 =$
$(x \cdot x)(x \cdot x \cdot x) =$
x^5

You can see that you have 5 x's, which is written as x^5. To get x^5 for an answer, you *add* the exponents instead of multiplying them.

INSIDE TRACK

WHAT DO YOU do if you see an expression like x^{20} and you want to multiply it by x^{25}? You can see that writing out the factors of $x^{20} \cdot x^{25}$ would take a long time. Think about how easy it would be to make a mistake if you wrote out all the factors. It is much more efficient and fast to use the rule for multiplying exponents: When you are multiplying, you add the exponents.

You can multiply variables only when the base is the same. You can multiply $a^2 \cdot a^4$ to get a^6. However, you cannot multiply $a^3 \cdot b^4$ because the bases are different.

To multiply $2x^2 \cdot 3x^4$, multiply the coefficients, keep the variable the same, and add the exponents. Your answer would be $6x^6$. When you multiply $5a^3 \cdot 2a^4$, you get $10a^7$.

PRACTICE LAP

DIRECTIONS: Use scratch paper to solve the following problems. You can check your answers at the end of this chapter.

Simplify the expressions using the rules for adding and multiplying with exponents.

1. $11x + 17x$

2. $3x + 4x^2$

3. $a^7 \cdot a^3$

4. $x^2y + 5x^2y$

5. $3x \cdot 4x^2$

6. $ab^2 \cdot a^2b^3$

AND THEN THERE WAS DIVISION

Now you know that when you multiply expressions with exponents, you add the exponents. However, when you divide, you *subtract* the exponents.

> **CAUTION!**
>
> **WHEN SUBTRACTING EXPONENTS**, always subtract the larger exponent from the smaller exponent. If the larger exponent is in the numerator, the variable and exponent will be in the numerator. For example, $\frac{x^5}{x^2} = x^3$. If the larger exponent is in the denominator, the variable and exponent will be in the denominator. For example, $\frac{x^2}{x^5} = \frac{1}{x^3}$.

Examples

$$\frac{x^5}{x^2} = \frac{x \cdot x \cdot x \cdot x \cdot x}{x \cdot x} = x^{(5-2)} = x^3$$

$$\frac{a^3 b^2}{a b^5} = \frac{a \cdot a \cdot a \cdot b \cdot b}{a \cdot b \cdot b \cdot b \cdot b \cdot b} = \frac{a^{(3-1)}}{b^{(5-2)}} = \frac{a^2}{b^3}$$

Writing out the factors when you work problems takes too long. So, you can use the rule for exponents when you divide. When you divide expressions that contain exponents like $\frac{4x^5}{2x^3}$, divide the coefficients (4 by 2) and subtract the exponents (5 minus 3). The answer for the expression is $2x^2$.

FUEL FOR THOUGHT

A QUANTITY RAISED to the zero power is 1. Example: $\frac{x^3}{x^3} = x^0$.
However, a number divided by itself is always 1. Therefore, x^0 must
equal 1.

PRACTICE LAP

DIRECTIONS: Use scratch paper to solve the following problems.
You can check your answers at the end of this chapter.
  Simplify the expressions.

7. $\frac{y^8}{y^3}$

8. $\frac{a^3}{a^6}$

9. $-\frac{b^7}{b^3}$

10. $\frac{a^4 b}{a^2}$

WHAT DO YOU DO WITH EXPONENTS WHEN YOU RAISE A QUANTITY TO A POWER?

How do you simplify the expression $(x^3)^2$? Remember that an exponent
tells you how many times a quantity is a factor.

Examples

$$(x^3)^2 = (x \cdot x \cdot x)(x \cdot x \cdot x) = x^6$$
$$(3a^3)^2 = (3 \cdot a \cdot a \cdot a)(3 \cdot a \cdot a \cdot a) = 3^2 a^6 = 9a^6$$

From the previous examples, you can see that if you multiply the exponents, you will get the correct answer. If you raise a quantity to a power, you
multiply the exponents. Note that if a number has no exponent, the exponent is understood to be 1.

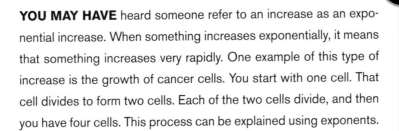

FUEL FOR THOUGHT

YOU MAY HAVE heard someone refer to an increase as an exponential increase. When something increases exponentially, it means that something increases very rapidly. One example of this type of increase is the growth of cancer cells. You start with one cell. That cell divides to form two cells. Each of the two cells divide, and then you have four cells. This process can be explained using exponents.

$2^0 = 1$

$2^1 = 2$

$2^2 = 4$

$2^3 = 8$

$2^4 = 16$

$2^5 = 32$

$2^6 = 64$

$2^7 = 128$

You can see the rapid increase, which explains why cancer can be so devastating.

What would be another real-world example of exponential growth?

RADICALS REIGN

You have seen how the addition in $x + 5 = 11$ can be undone by subtracting 5 from both sides of the equation. You have also seen how the multiplication in $3x = 21$ can be undone by dividing both sides by 3. Taking the **square root** (also called a **radical**) is the way to undo the exponent from an equation like $x^2 = 25$.

The exponent in 7^2 tells you to square 7. You multiply $7 \cdot 7$ and get $7^2 = 49$.

The **radical sign** $\sqrt{}$ in $\sqrt{36}$ tells you to find the positive number whose square is 36. In other words, $\sqrt{36}$ asks, what positive number times itself is 36? The answer is $\sqrt{36} = 6$ because $6 \cdot 6 = 36$.

The number inside the radical sign is called the **radicand**. For example, in $\sqrt{9}$, the radicand is 9.

Square Roots of Perfect Squares

The easiest radicands to deal with are perfect squares. Because they appear so often, it is useful to learn to recognize the first few perfect squares: $0^2 = 0$, $1^2 = 1$, $2^2 = 4$, $3^2 = 9$, $4^2 = 16$, $5^2 = 25$, $6^2 = 36$, $7^2 = 49$, $8^2 = 64$, $9^2 = 81$, $10^2 = 100$, $11^2 = 121$, and $12^2 = 144$.

It is even easier to recognize when a variable is a perfect square because the exponent is even. For example, $x^{14} = x^7 \cdot x^7$ and $a^8 = a^4 \cdot a^4$.

Example
$$\sqrt{64x^2y^{10}}$$

Write as a square: $\sqrt{8xy^5 \cdot 8xy^5}$

Evaluate: $8xy^5$

You could have also split the radical into parts and evaluated them separately:

Example
$$\sqrt{64x^2y^{10}}$$

Split into perfect squares: $\sqrt{64 \cdot x^2 \cdot y^{10}}$

Write as squares: $\sqrt{8 \cdot 8} \cdot \sqrt{x \cdot x} \cdot \sqrt{y^5 \cdot y^5}$

Evaluate: $8 \cdot x \cdot y^5$

Multiply: $8xy^5$

INSIDE TRACK

IF YOUR RADICAL has a coefficient like $3\sqrt{25}$, evaluate the square root before multiplying: $3\sqrt{25} = 3 \cdot 5 = 15$.

PRACTICE LAP

DIRECTIONS: Use scratch paper to solve the following problems. You can check your answers at the end of this chapter.

11. $\sqrt{49}$

12. $\sqrt{81}$

13. $\sqrt{144}$

14. $-\sqrt{64}$

15. $4\sqrt{4}$

16. $-2\sqrt{9}$

17. $\sqrt{a^2}$

18. $5\sqrt{36}$

Simplifying Radicals

Not all radicands are perfect squares. There is no whole number that, when multiplied by itself, equals 5. With a calculator, you can get a decimal that squares very close to 5, but it won't come out exactly. The only precise way to represent the square root of 5 is to write $\sqrt{5}$. It cannot be simplified any further.

There are three rules for knowing when a radical can be simplified further:

1. The radicand contains a factor, other than 1, that is a perfect square.
2. The radicand is a fraction.
3. The radical is in the denominator of a fraction.

When the Radicand Contains a Factor That Is a Perfect Square

To determine if a radicand contains any factors that are perfect squares, factor the radicand completely. All the factors must be prime. A number is prime if its only factors are 1 and the number itself. A prime number cannot be factored any further.

For example, here's how you simplify $\sqrt{12}$. The number 12 can be factored into $12 = 2 \cdot 6$. This is not completely factored because 6 is not prime. The number 6 can be further factored $6 = 2 \cdot 3$. The number 12 completely factored is $2 \cdot 2 \cdot 3$.

The radical $\sqrt{12}$ can be written as $\sqrt{2 \cdot 2 \cdot 3}$. This can be split up into $\sqrt{2 \cdot 2} \cdot \sqrt{3}$. Because $\sqrt{2 \cdot 2} = 2$, the simplified form of $\sqrt{12}$ is $2\sqrt{3}$.

Example

$\sqrt{18}$

Factor completely:	$\sqrt{2 \cdot 3 \cdot 3}$
Separate out the perfect square $3 \cdot 3$:	$\sqrt{3 \cdot 3} \cdot \sqrt{2}$
Simplify:	$3\sqrt{2}$

Example

$\sqrt{32}$

Factor completely:	$\sqrt{2 \cdot 16}$
The number 16 is not prime. It can be factored:	$\sqrt{2 \cdot 2 \cdot 8}$
The number 8 is not prime. It can be factored:	$\sqrt{2 \cdot 2 \cdot 2 \cdot 4}$
The number 4 is not prime. It can be factored:	$\sqrt{2 \cdot 2 \cdot 2 \cdot 2 \cdot 2}$

You have two sets of perfect squares, $2 \cdot 2$ and $2 \cdot 2$. The square root of each is 2, so you have two 2's outside the radical. You then multiply the numbers that are outside the radical: $2 \cdot 2\sqrt{2}$
Simplify. The product of 2 times 2 gives you 4: $4\sqrt{2}$

PRACTICE LAP

DIRECTIONS: Use scratch paper to solve the following problems. You can check your answers at the end of this chapter.

19. $\sqrt{8}$
20. $\sqrt{20}$
21. $\sqrt{54}$
22. $\sqrt{40}$
23. $\sqrt{72}$
24. $\sqrt{27}$

When the Radicand Contains a Fraction

The radicand cannot be a fraction. If you get rid of the denominator in the radicand, then you no longer have a fraction. This process is called rationalizing the denominator. Your strategy will be to make the denominator a perfect square. To do that, you multiply the denominator by itself. However, if you multiply the denominator of a fraction by a number, you must multiply the numerator of the fraction by the same number.

Example

$$\sqrt{\frac{1}{2}}$$

Make the denominator a perfect square: $\sqrt{\frac{1}{2} \cdot \frac{2}{2}}$
Take out the square roots.
One is a perfect square and so is $2 \cdot 2$: $\frac{1}{2}\sqrt{2}$

Now let's look at another example.

Example

$$\sqrt{\frac{2}{3}}$$

Make the denominator a perfect square: $\sqrt{\frac{2}{3} \cdot \frac{3}{3}}$
The number 1 is considered a factor of all numbers. If the numerator does not contain a perfect square, then 1 will be the perfect square and will be in the numerator. Take the square root of 1 in the numerator and $3 \cdot 3$ in the denominator. The product of $2 \cdot 3$ will give you 6 for the radicand. Your final answer will be $\frac{1}{3}\sqrt{6}$.

PRACTICE LAP -

DIRECTIONS: Use scratch paper to solve the following problems. You can check your answers at the end of this chapter.

25. $\sqrt{\frac{2}{5}}$

26. $\sqrt{\frac{2x^2}{3}}$

27. $\sqrt{\frac{a^2b^2}{2}}$

When There Is a Radical in the Denominator

When you have a radical in the denominator, the expression is not in simplest form. The expression $\frac{2}{\sqrt{3}}$ contains a radical in the denominator. To get rid of the radical in the denominator, rationalize the denominator. In other words, make the denominator a perfect square. To do that, you need to multiply the denominator by itself:

$$\frac{2}{\sqrt{3}} \cdot \frac{\sqrt{3}}{\sqrt{3}}$$

Now, simplify $\frac{2\sqrt{3}}{\sqrt{9}}$. The number 9 is a perfect square.

PRACTICE LAP -

DIRECTIONS: Use scratch paper to solve the following problems. You can check your answers at the end of this chapter.

28. $\frac{3}{\sqrt{7}}$

29. $\frac{6}{\sqrt{5}}$

ADDING AND SUBTRACTING SQUARE ROOTS

Square roots are easy to add or subtract. You can add or subtract radicals if the radicands are the same. To add or subtract radicals, you add the number in front of the radicals and leave the radicand the same. When you add $3\sqrt{2}$ and $5\sqrt{2}$, you add the 3 and the 5, but the radicand $\sqrt{2}$ stays the same. The answer is $8\sqrt{2}$.

CAUTION!

YOU CAN ADD or subtract radicals *only* when the radicand is the same. You add radicals by adding the number in front of the radicals and keeping the radicand the same. When you subtract radicals, you subtract the numbers in front of the radicals and keep the radicand the same.

PRACTICE LAP

DIRECTIONS: Use scratch paper to solve the following problems. You can check your answers at the end of this chapter.

30. $3\sqrt{7} + 8\sqrt{7}$
31. $11\sqrt{3} - 8\sqrt{3}$
32. $5\sqrt{2} + 6\sqrt{2} - 3\sqrt{2}$

MULTIPLYING AND DIVIDING RADICALS

It is easy to multiply and divide radicals. To multiply radicals like $4\sqrt{3}$ and $2\sqrt{2}$, you multiply the numbers in front of the radicals: 4 times 2. Then, multiply the radicands: $\sqrt{3}$ times $\sqrt{2}$. The answer is $8\sqrt{6}$.

Example

$$5\sqrt{3} \cdot 2\sqrt{2}$$

Multiply the numbers in front of the radicals. Then, multiply the radicands. You will end up with $10\sqrt{6}$.

To divide the radical $4\sqrt{6}$ by $2\sqrt{3}$, divide the numbers in front of the radicals. Then, divide the radicands. The answer is $2\sqrt{2}$.

INSIDE TRACK

WHEN YOU MULTIPLY or divide radicals, the radicands do not have to be the same.

PRACTICE LAP

DIRECTIONS: Use scratch paper to solve the following problems. You can check your answers at the end of this chapter.

33. $7\sqrt{3} \cdot 5\sqrt{2}$

34. $\frac{14\sqrt{6}}{7\sqrt{2}}$

35. $-3\sqrt{5} \cdot 4\sqrt{2}$

SOLVING RADICAL EQUATIONS

A radical equation contains a variable in the radicand, like $\sqrt{x} = 3$. You know that squaring something is the opposite of taking the square root. To solve a radical equation, you square both sides, getting rid of the square roots. **Note:** This is true only if the radical is a square root.

Example

$\sqrt{x} = 3$

Square both sides:	$(\sqrt{x})^2 = 3^2$
Simplify:	$\sqrt{x} \cdot \sqrt{x} = 3 \cdot 3$
Multiply:	$\sqrt{x \cdot x} = 9$
Take the square root of x^2:	$x = 9$

To solve an equation like $x^2 = 25$ requires a little extra thought. Plug in $x = 5$ and you see that $5^2 = 25$. This means that $x = 5$ is a solution to $x^2 = 25$. However, if you plug in $x = -5$, you see that $(-5)^2 = (-5) \cdot (-5) = 25$ also. This means that $x = -5$ is also a solution to $x^2 = 25$. The equation $x^2 = 25$ has two solutions: $x = 5$ and $x = -5$. This happens so often that there is a special symbol $\pm$ that means *plus or minus*. You say that $x = \pm 5$ is the solution to $x^2 = 25$.

Now that you know what a radical equation is, look at how to solve radical equations that require more than one step.

Example

$3\sqrt{x} = 15$

Divide both sides of the equation by 3:	$\frac{(3\sqrt{x})}{3} = \frac{15}{3}$
Simplify:	$\sqrt{x} = 5$
Square both sides of the equation:	$(\sqrt{x})^2 = 5^2$
Simplify:	$x = 25$

Just for good measure, take a look at another example.

Example

$2\sqrt{x + 2} = 10$

Divide both sides of the equation by 2:	$\frac{(2\sqrt{x+2})}{2} = \frac{10}{2}$
Simplify:	$\sqrt{x + 2} = 5$
Square both sides of the equation:	$(\sqrt{x + 2})^2 = 5^2$
Simplify:	$x + 2 = 25$
Subtract 2 from both sides of the equation:	$x + 2 - 2 = 25 - 2$
Simplify:	$x = 23$

FUEL FOR THOUGHT

WHEN YOU MULTIPLY a radical by itself, the radical sign disappears.

$\sqrt{x} \cdot \sqrt{x} = x$ and $\sqrt{x+3} \cdot \sqrt{x+3} = x+3$

PRACTICE LAP

DIRECTIONS: Use scratch paper to solve the following problems. You can check your answers at the end of this chapter.

36. $x^2 = 81$

37. $\sqrt{n} = 3 \cdot 2$

38. $\sqrt{x} + 5 = 7$

39. $4\sqrt{x} = 20$

40. $3\sqrt{x} + 2 = 8$

ANSWERS

1. $28x$

2. $3x + 4x^2$

3. a^{10}

4. $6x^2y$

5. $12x^3$

6. a^3b^5

7. y^5

8. $\frac{1}{a^3}$

9. $-b^4$

10. a^2b

11. 7

12. 9

13. 12

14. -8

15. 8

16. -6

17. a

18. 30

19. $2\sqrt{2}$

20. $2\sqrt{5}$

21. $3\sqrt{6}$

22. $2\sqrt{10}$

23. $6\sqrt{2}$

24. $3\sqrt{3}$

25. $\frac{1}{5}\sqrt{10}$

26. $\frac{x\sqrt{6}}{3}$

27. $\frac{ab\sqrt{2}}{2}$

28. $\frac{3\sqrt{7}}{7}$

29. $\frac{6\sqrt{5}}{5}$

30. $11\sqrt{7}$

31. $3\sqrt{3}$

32. $8\sqrt{2}$

33. $35\sqrt{6}$

34. $2\sqrt{3}$

35. $-12\sqrt{10}$

36. ± 9

37. 36

38. 4

39. 25

40. 4

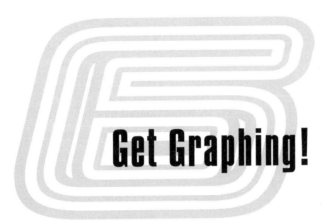

Get Graphing!

WHAT'S AROUND THE BEND

- → Graphing Points
- → Slope
- → *Y*-Intercept
- → Slope-Intercept Form
- → Graphing Linear Equations
- → Word Problems

A graph is a picture. When you graph an equation, you are creating a picture of your answer. To graph equations with two variables, you use a coordinate plane. Two intersecting lines that meet at right angles form a **coordinate plane**.

The two intersecting lines that form the coordinate plane intersect at a point called the **origin**. The origin is your starting point. The horizontal line is called the **x-axis**. When you move to the right of the origin on the *x*-axis, the numbers are positive. When you move to the left of the origin on the *x*-axis, the numbers are negative.

The vertical axis is called the **y-axis**. When you move above the origin on the y-axis, the numbers are positive. When you move below the origin, the numbers are negative.

The x- and y-axes divide the plane into four equal parts. These parts are called **quadrants** and are named by a number. The quadrant in the upper right-hand corner is quadrant 1. To label the other quadrants, go counterclockwise.

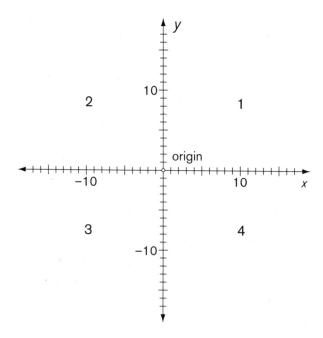

PLOTTING POINTS ON A GRAPH

You can plot points on a coordinate plane. Every point has two coordinates: an x-coordinate and a y-coordinate. These coordinates are written as an ordered pair. An **ordered pair** is a pair of numbers with a special order. The pair of numbers is enclosed in parentheses with the x-coordinate first and the y-coordinate second. For example, the ordered pair (2,3) has an x-coordinate of 2 and a y-coordinate of 3. To plot this point, start at the origin and move two units in the positive direction on the x-axis. From there, move up three units because the y-coordinate is a positive 3.

To plot the ordered pair (4,2), you start at the origin and move four units in the positive direction on the x-axis. From there, move two units up because 2 is positive. To graph the point (−5,3), you start at the origin and move five units in the negative direction on the *x*-axis. From there, move up three units. To graph $(-2,-4)$, you start at the origin and move two units in the negative direction on the *x*-axis. From that point, move down four units because the 4 is negative.

Example

Look at the points on the coordinate plane that follows. Each letter names an ordered pair. The ordered pairs are listed below the coordinate plane.

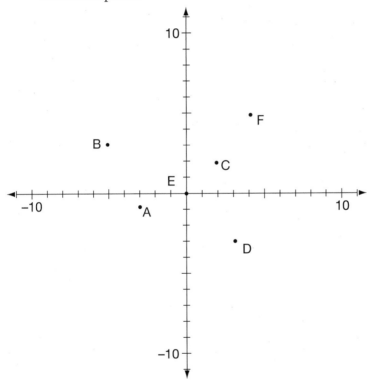

A $(-3,-1)$
B $(-5,3)$
C $(2,2)$
D $(3,-3)$
E $(0,0)$
F $(4,5)$

DIRECTIONS: Use graph paper to plot the following points. You can plot all these points on the same coordinate plane.

1. (8,2)
2. (8,−2)
3. (0,4)
4. (5,0)
5. (−5,−5)
6. (−1,5)
7. (6,−3)
8. (−6,−2)
9. (2,9)
10. (−7,3)

USING THE SLOPE AND *Y*-INTERCEPT

Did you find plotting points easy? If so, you are ready to move on and graph linear equations, but first, you need to know what the terms *slope* and *y-intercept* mean. There are several methods that can be used to graph linear equations; however, you will use the slope-intercept method here.

What does slope mean to you? If you are a skier, you may think of a ski slope. The slope of a line has a similar meaning. The **slope** of a line is the steepness of a line. What is the *y*-intercept? Intercept means to interrupt, so the **y-intercept** is where the line interrupts, or runs through, the *y*-axis.

To graph a linear equation, you will first change its equation into slope-intercept form. The slope-intercept form of a linear equation is $y = mx + b$, also known as *y = form*. Linear equations have two variables. For example, in the equation $y = mx + b$, the two variables are x and y. The m represents a number and is the slope of the line. The b represents a number and is the *y*-intercept. For example, in the equation $y = 2x + 3$, the number 2 is the m, which is the slope. The 3 is the b, which is the *y*-intercept. In the equation $y = -3x + 5$, m or the slope is −3, and b or the *y*-intercept is 5.

DIRECTIONS: Use scratch paper to solve the following problems. You can check your answers at the end of this chapter.

Find the slope (*m*) and *y*-intercept (*b*) of each equation.

11. $y = 3x + 9$
12. $y = 5x - 6$
13. $y = -5x + 16$
14. $y = -2.3x - 7.5$
15. $y = \frac{3}{4}x + 5$
16. $y = \frac{x}{3} + 8$ (Hint: The number in front of the *x* is 1.)

Getting the Right Form

What if the equation is not in slope-intercept form? Simple! All you need to do is change the equation to slope-intercept form. How? Slope-intercept form is *y = form*, so your strategy is to get the *y* on a side by itself.

An equation needs to be in slope-intercept form or *y = form* ($y = mx + b$) before you can graph the equation with a pencil and graph paper. Also, if you were to use a graphing calculator to graph a linear equation, the equation would need to be in *y = form* before it can be entered into the calculator.

Example

$$2x + y = 5$$

Subtract 2*x* from both sides of the equation: $2x - 2x + y = 5 - 2x$
Simplify: $y = 5 - 2x$
Rearrange the equation so the *x* term is first: $y = -2x + 5$

A mathematical rule called the **commutative property** lets you change the order of numbers or terms when you add or multiply. You want the previous equation in the form $y = mx + b$, so the order of the 5 and the −2*x* needed to be changed after getting the *y* on a side by itself. When you move a term, be sure to take the sign of the term with it. For example, the 5 was a positive number. It remains a positive number when you move it.

Example

$2x + 3y = 9$

Subtract $2x$ from both sides of the equation:	$2x - 2x + 3y = 9 - 2x$
Simplify:	$3y = 9 - 2x$
Use the commutative property:	$3y = -2x + 9$
Divide both sides by 3:	$y = -\frac{2x}{3} + \frac{9}{3}$
Simplify both sides of the equation:	$y = -\frac{2}{3}x + 3$

PRACTICE LAP

DIRECTIONS: Use scratch paper to solve the following problems. You can check your answers at the end of this chapter.

Change the equations to slope-intercept form. State the slope as *m* and the *y*-intercept as *b*.

17. $2x + y = -4$

18. $3x + y = 6$

19. $-3x + y = 8$

20. $5x + 5y = 15$

21. $-20x + 10y = 50$

Graphing Linear Equations Using the Slope and *Y*-Intercept

The graph of a linear equation is a line, which means it goes on forever in both directions. A graph is a picture of all the answers to the equation, so there is an infinite (endless) number of solutions. Every point on that line is a solution.

CAUTION!

LOOK AT ALL coordinate planes carefully to see what scale each is drawn to—some may have increments of one, while others may have increments of one-half or ten.

You know that slope means the steepness of a line. In the equation $y = 2x + 3$, the slope of the line is 2. What does it mean when you have a slope of 2? Slope is defined as the rise of the line over the run of the line. If the slope is 2, this means $\frac{2}{1}$, so the rise is 2 and the run is 1.

If the slope of a line is $\frac{2}{3}$, the rise is 2 and the run is 3. What do rise and run mean? **Rise** is the vertical change, and **run** is the horizontal change. To graph a line passing through the origin with a slope of $\frac{2}{3}$, start at the origin. The rise is 2, so from the origin, go up two and to the right three. Then, draw a line from the origin to the endpoint. The line you have drawn has a slope of $\frac{2}{3}$.

Now draw a line with a slope of $-\frac{3}{4}$. Start at the origin. Go down three units because you have a negative slope. Then, go right four units. Finally, draw a line from the origin to the endpoint. These two lines appear on the same graph that follows.

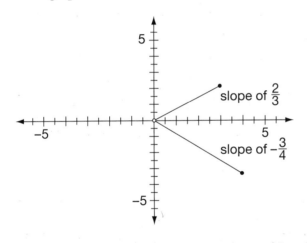

To graph an equation like $y = x + 1$, you can use the slope and the y-intercept. The first step is to determine the slope. The slope is the number in front of x, which means, in this case, it is 1. What is the y-intercept? It is also 1. To graph the equation, your starting point will be the y-intercept, which is 1. From the y-intercept, use the slope, which is also 1, or $\frac{1}{1}$. The slope tells you to go up one and to the right one. A line is drawn from the y-intercept to the point $(1,2)$. You can extend this line and draw arrows on each end to show that the line extends infinitely.

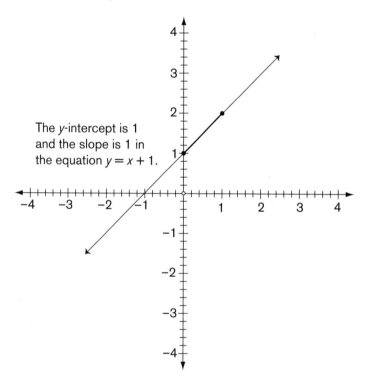

The y-intercept is 1 and the slope is 1 in the equation $y = x + 1$.

Example

Graph the equation $y = -\frac{2}{3}x + 2$. Start with the y-intercept, which is a positive 2. From there, go down two (because the slope is negative) and to the right three. Draw a line to connect the intercept and the endpoint.

In the equation $y = -\frac{2}{3}x + 2$, the slope is $-\frac{2}{3}$ and the y-intercept is 2.

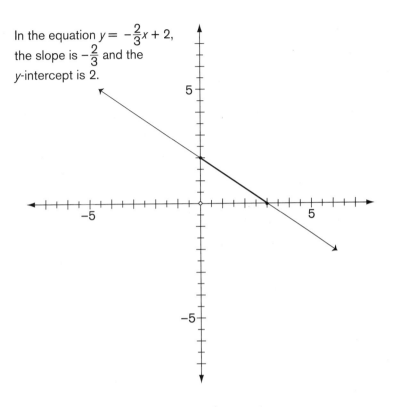

Two Special Types of Lines

Any equation in the form $y = k$ (where k is a constant), such as $y = 3$ or $y = -2$, will be a horizontal line. The slope of the horizontal line will be zero, because there is no change in the y-values. In other words, the slope formula will have zero in the numerator, which makes the entire fraction equal to zero. The equation of the x-axis is $y = 0$.

INSIDE TRACK

TO REMEMBER THAT the slope of a vertical line is zero, think about traveling in a straight line. The line is flat, so the slope, or steepness, would be zero.

Any equation in the form $x = k$ (where k is a constant), such as $x = 3$ or $x = -2$, will be a vertical line. The line will have an undefined slope.

Applications

Writing equations in the slope-intercept form of a linear equation, $y = mx + b$, can be useful in solving word problems. Additionally, graphing linear equations can help solve word problems. Word problems are an important part of algebra.

Try this one. If an airplane maintains a landing approach of a constant rate of descent of 50 feet for every 500 feet horizontally, what is the slope of the line that represents the plane's landing approach?

The rise of the landing approach is down 50, which would be represented by –50. The run of the landing approach is 500. The slope is represented by the rise over the run, which is $-\frac{50}{500} = -\frac{1}{10}$.

Just for good measure, here's another word problem. You are charged a flat fee of $5 a month plus $0.11 per kilowatt-hour of power used. Write an equation that would calculate your power bill for a month. Then state the slope and y-intercept of your equation.

You will be charged $0.11 for each kilowatt-hour of power you use, so let x equal the number of kilowatt-hours. Let y equal the monthly power bill. Your equation will be $y = 0.11x + 5$. The slope of the equation is 0.11, and the y-intercept is 5.

DIRECTIONS: Use scratch paper to solve the following problems. You can check your answers at the end of this chapter.

22. An airplane maintains a landing approach of a constant rate of descent of 70 feet for every 700 feet traveled horizontally. What is the slope of the line that represents the plane's landing approach?

23. You are a sales clerk in a clothing store. You receive a salary of $320 a week plus a 5% commission on all sales. Write an equation to represent your weekly salary. (Let y equal the weekly salary and x equal the amount of sales.) What is the slope and y-intercept of your equation?

24. You are renting a car from Economy Rent-A-Car. It will cost you $25 a day plus $0.10 a mile per day. Write an equation to represent the daily cost of renting the car. Let y equal the daily cost and let x equal the number of miles traveled per day. What is the slope and y-intercept of your equation?

ANSWERS

1–10.

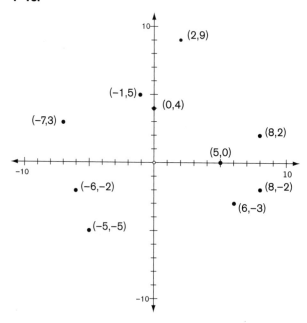

11. $m = 3, b = 9$

12. $m = 5, b = -6$

13. $m = -5, b = 16$

14. $m = -2.3, b = -7.5$

15. $m = \frac{3}{4}, b = 5$

16. $m = \frac{1}{3}, b = 8$

17. $y = -2x - 4, m = -2, b = -4$

18. $y = -3x + 6, m = -3, b = 6$

19. $y = 3x + 8, m = 3, b = 8$

20. $y = -x + 3, m = -1, b = 3$

21. $y = 2x + 5, m = 2, b = 5$

22. $-\frac{70}{700} = -\frac{1}{10}$

23. $y = 5\% \cdot x + 320,$

$m = \frac{5}{100} = \frac{1}{20}, b = 320$

24. $y = 0.10x + 25, m = 0.10, b = 25$

Systems of Equations

WHAT'S AROUND THE BEND

➤ Solving a System of Equations
➤ Parallel Lines
➤ Perpendicular Lines
➤ Negative Reciprocals
➤ Coincident Lines
➤ Solving Systems Algebraically

A **system of equations** is a group of two or more equations. In order to solve a system of equations by graphing, each equation is graphed and then the point of intersection of the lines is identified. This point is the place where the two or more equations are equal to each other; the point is the solution to the system of equations.

Example

Solve the following system of equations by graphing.

$y = 2x + 5$

$x + y = 2$

First, make sure that each equation is in slope-intercept ($y = mx + b$) form. This technique was introduced in Chapter 6. The first equation is already in the correct form. In the equation $y = 2x + 5$, the slope (m) is 2 and the y-intercept (b) is 5.

The second equation needs to be transposed to the correct form. To do this, subtract x from both sides of the equation: $x + y = 2$ becomes $x - x + y = 2 - x$.

Write the x term first on the right side of the equation: $y = -x + 2$.

This equation is now in slope-intercept form. The slope of the line is –1, and the y-intercept is 2.

To solve this system of equations graphically, graph both equations on the same set of axes and look for where the two lines cross, or intersect, each other. This graph shows the system of $y = 2x + 5$ and $x + y = 2$.

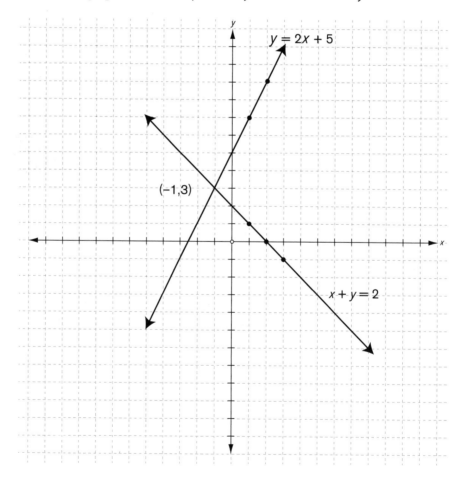

Because the two lines cross at the point (–1,3), the solution to the system is at the point where $x = -1$ and $y = 3$.

To check the solution to the system, substitute the x- and y-value of the solution point into **both** equations.

$$y = 2x + 5 \qquad\qquad x + y = 2$$
$$3 = 2(-1) + 5 \qquad -1 + 3 = 2$$
$$3 = 3 \qquad\qquad\qquad 2 = 2$$

Because both equations are true, the solution is correct.

THREE SPECIAL CASES OF SYSTEMS

There are three special cases of linear equations that you may encounter. The lines may be **parallel** or **perpendicular**, or they may **coincide** with each other. The first case is parallel lines.

FUEL FOR THOUGHT

PARALLEL LINES ARE lines in the same plane that have the same slope and will never intersect.

Parallel lines have the same slope. When solving graphically, you may notice that the values for m are the same, even before you graph the equations.

Example

Solve the following system of equations by graphing.

$$x + y = 4$$
$$y = -x - 3$$

First, put each equation into slope-intercept form.

Subtract x from both sides of the first equation and write the x term first on the right side.

$$x - x + y = -x + 4$$
$$y = -x + 4$$

The slope is –1, and the y-intercept is 4.
The second equation is already in the correct form.

$$y = -x - 3$$

The slope is –1, and the y-intercept is –3.
Notice that both equations have a slope of –1, but each has a different y-intercept.
Graph both equations on the same set of axes.

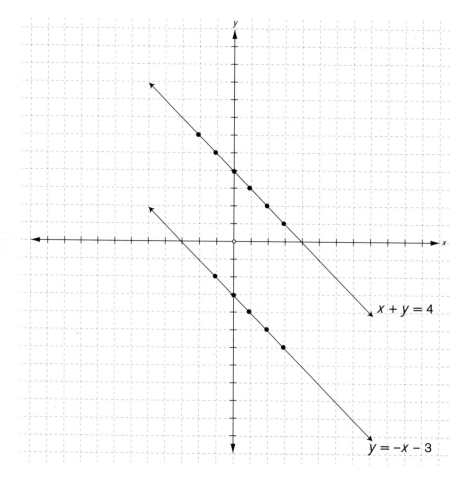

These two lines run next to each other, or are parallel, and will never cross because the slope of each line is the same. There is no solution to this type of system.

The second case of linear equations is perpendicular lines.

FUEL FOR THOUGHT

PERPENDICULAR LINES ARE lines that meet to form right angles. Their slopes are negative reciprocals of each other. **Negative reciprocals** are two fractions whereby one is positive and the other is negative, and the numerators and denominators are switched. Some pairs of negative reciprocals are $\frac{1}{4}$ and -4, $\frac{-5}{6}$ and $\frac{6}{5}$, and 1 and -1.

Example

Solve the following system of equations by graphing.

$$y = \frac{3}{2}x - 1$$
$$y = \frac{-2}{3}x - 1$$

Both equations are in the slope-intercept form.
In the first equation, the slope is $\frac{3}{2}$ and the y-intercept is -1.
In the second equation, the slope is $\frac{-2}{3}$ and the y-intercept is -1.
Notice that the slopes are negative reciprocals of each other: $\frac{3}{2}$ and $\frac{-2}{3}$.
Graph the two equations on the same set of axes to find the solution.

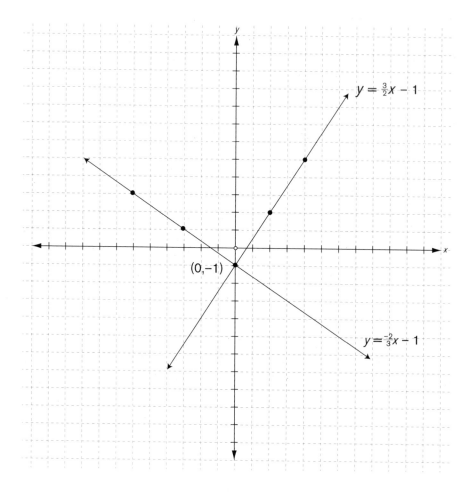

The two lines intersect at the point $(0,-1)$, so this is the solution to the system. These two lines meet to form right angles and will have exactly one solution.

The third case of linear equations involves lines that coincide with each other. **Coincident lines** are two lines that have the same equation. On a graph, they represent the same line. The solution set to this type of system is *all* of the points on the line, or an infinite number of solutions.

FUEL FOR THOUGHT

COINCIDENT LINES ARE two lines that have the same equation; on a graph, they represent the same line.

When solving a system of equations that coincide, you will notice something unique as soon as you begin to graph the lines. Each of the equations, although they may appear different from each other at the start, represents the same line.

Example

Solve the following system of equations by graphing.

$$y - 5 = 3x$$
$$2y - 6x = 10$$

Change each equation to slope-intercept form.

Add 5 to both sides of the first equation: $y - 5 + 5 = 3x + 5$. This simplifies to $y = 3x + 5$.

The slope of the line is 3, and the y-intercept is 5.

For the second equation, add $6x$ to both sides: $2y - 6x + 6x = 10 + 6x$. This simplifies to $2y = 10 + 6x$.

Divide both sides by 2 and write the x term first on the right side of the equation:

$$\frac{2y}{2} = \frac{6x}{2} + \frac{10}{2}$$

This simplifies to $y = 3x + 5$. The slope of this line is 3, and the y-intercept is 5. Graph both equations on the same set of axes.

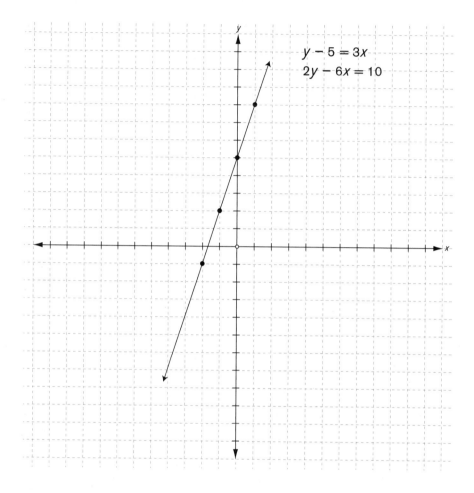

$$y - 5 = 3x$$
$$2y - 6x = 10$$

When graphed on a set of axes, the lines are exactly on top of each other. The graph of each equation is the same line. In this special case, both equations share *all* solutions. Therefore, the solution set to this type of system is all points on the line, or an infinite number of solutions.

INSIDE TRACK

THE SOLUTION TO a linear system of equations will have one solution, no solution, or an infinite number of solutions.

SOLVING SYSTEMS OF EQUATIONS ALGEBRAICALLY

When solving a system of equations, you are finding the place or places where two or more equations equal each other. There are two ways to do this algebraically: by elimination and by substitution.

Elimination Method

In this method, you will be using addition or subtraction to eliminate one of the two variables so that the equation you are working with has only one variable.

Solve the system $x - y = 6$ and $2x + 3y = 7$.

Put the equations one above the other, lining up the x's, y's, and the equal sign.

$$x - y = 6$$
$$2x + 3y = 7$$

Multiply the first equation by -2 so that the coefficients of x are opposites. This will allow the x's to cancel out in the next step. Make sure that *all* terms are multiplied by -2. The second equation remains the same.

$$-2(x - y = 6) \rightarrow -2x + 2y = -12$$
$$2x + 3y = 7 \rightarrow 2x + 3y = 7$$

Combine the new equations vertically.

$$-2x + 2y = -12$$
$$\underline{2x + 3y = 7}$$
$$5y = -5$$

Divide both sides by 5.

$$\frac{5y}{5} = \frac{-5}{5}$$
$$y = -1$$

To complete the problem, solve for *x* by substituting –1 for *y* into one of the original equations.

$$x - (-1) = 6$$
$$x + 1 = 6$$
$$x + 1 - 1 = 6 - 1$$
$$x = 5$$

The solution to the system is *x* = 5 and *y* = –1, or (5,–1).

CAUTION!

ALWAYS LINE UP like terms when using the elimination method.

PRACTICE LAP

DIRECTIONS: Use scratch paper to solve the following problems. You can check your answers at the end of this chapter.

Solve the systems of equations using the elimination method.

1. $x + y = 8$
$x - y = 18$

2. $x - 2y = 4$
$x + 2y = 16$

3. $-3x + 2y = -12$
$3x + 5y = -9$

4. $x + y = 13$
$x + 2y = 1$

5. $2x - y = 7$
$4x - y = 9$

6. $3x + 2y = 10$
$4x + 2y = 2$

PRACTICE LAP

7. $2x + 3y = 23$

  $7x - 3y = 13$

8. $8x - 4y = 16$

  $4x + 5y = 22$

9. $5x - 3y = 31$

  $2x + 5y = 0$

Substitution Method

In this method, you will be substituting a quantity for one of the variables to create an equation with only one variable.

Solve the system $x + 2y = 5$ and $y = -2x + 7$.

Substitute the second equation into the first for y.

$$x + 2(-2x + 7) = 5$$

Use distributive property to remove the parentheses.

$$x + -4x + 14 = 5$$

Combine like terms. Remember that $x = 1x$.

$$-3x + 14 = 5$$

Subtract 14 from both sides and then divide by −3.

$$-3x + 14 - 14 = 5 - 14$$
$$\frac{-3x}{-3} = \frac{-9}{-3}$$
$$x = 3$$

To complete the problem, solve for *y* by substituting 3 for *x* in one of the original equations.

$$y = -2(3) + 7$$
$$y = -6 + 7$$
$$y = 1$$

The solution to the system is $x = 3$ and $y = 1$, or (3,1).

INSIDE TRACK

USE THE SUBSTITUTION METHOD when one of the equations already has either *x* or *y* alone.

PRACTICE LAP

DIRECTIONS: Use scratch paper to solve the following problems. You can check your answers at the end of this chapter.

Solve the systems of equations using the substitution method.

10. $x = 3y$
$2x + y = 14$

11. $x = -5y$
$2x + 2y = 16$

12. $y = 4x$
$3x + 4y = 38$

13. $y = 2x + 1$
$3x + 2y = 9$

14. $x = 2y + 1$
$3x - y = 13$

15. $y = 3x + 2$
$2x - 3y = 8$

ANSWERS

1. $(13,-5)$
2. $(10,3)$
3. $(2,-3)$
4. $(25,-12)$
5. $(1,-5)$
6. $(-8,17)$
7. $(4,5)$
8. $(3,2)$

9. $(5,-2)$
10. $(6,2)$
11. $(10,-2)$
12. $(2,8)$
13. $(1,3)$
14. $(5,2)$
15. $(-2,-4)$

More about Inequalities

WHAT'S AROUND THE BEND

➡ Solving Systems of Inequalities with One Variable
➡ Horizontal Lines
➡ Vertical Lines
➡ Solving Systems of Inequalities with Two Variables

You've already worked with inequalities. You should recall that the four symbols used when solving inequalities are

< is less than
> is greater than
≤ is less than or equal to
≥ is greater than or equal to

As a review, here are a few examples of simple inequalities:

$4 > 3$ is read "four is greater than three."
$-2 \le -1$ is read "negative two is less than or equal to negative one."

Solving an inequality is very similar to solving an equation. Most of the steps to solving inequalities are the same steps you use to solve equations. You perform the *inverse* operation of what you want to eliminate on both sides of the inequality. For example, in order to solve the inequality $5x \geq 25$, treat the *greater than or equal to* symbol like an equal sign. Divide both sides by 5 to get the *x* alone: $\frac{5x}{5} \geq \frac{25}{5}$. This simplifies to $x \geq 5$. For the solution, *x* is greater than or equal to 5. This means that *x* could be any number 5 or greater.

The major difference between an equation and an inequality is the symbol used; most inequalities imply that more than one value of the variable will make the mathematical statement true. Therefore, when you are solving an inequality with one variable, the solution set is graphed on a number line.

To graph a solution set, use the number in the solution as the starting point on the number line. In the problem $x \geq 5$, five is the starting point on the number line.

When using the symbol < or >, make an open circle at this number to show this number is not part of the solution set. When using the symbol ≤ or ≥, place a closed, or filled-in, circle at this number to show this number is a part of the solution set. In the case of $x \geq 5$, the circle is closed.

Next, draw an arrow from that point either to the left or the right on the number line. For the inequality $x \geq 5$, solutions to this problem are greater than or equal to five, so the line should be drawn to the right.

$$\xleftarrow{\quad} -1 \quad 0 \quad 1 \quad 2 \quad 3 \quad 4 \quad 5 \quad 6 \quad 7 \quad 8 \xrightarrow{\quad}$$

When you are solving inequalities, there is one catch. If you are multiplying or dividing each side of the inequality by a negative number, you must reverse the direction of the inequality symbol. Be careful to watch for this type of situation.

Example

Solve the inequality: $-3x + 6 \leq 18$

First, subtract 6 from both sides: $-3x + 6 - 6 \leq 18 - 6$

Then, divide both sides by −3 and
reverse the inequality symbol: $\frac{-3x}{-3} \geq \frac{12}{-3}$

Simplify: $x \geq -4$

The graph of this solution is

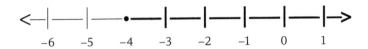

To check a solution, choose a number in the solution set and substitute into the original inequality. To check the inequality $-3x + 6 \leq 18$, check a value in the solution such as 0:

$$-3(0) + 6 \leq 18$$
$$0 + 6 \leq 18$$
$$6 \leq 18$$

Because 6 is less than or equal to 18, this solution is true.

PRACTICE LAP

DIRECTIONS: Use scratch paper to solve the following inequalities. You can check your answers at the end of this chapter.

1. $x + 3 < 10$
2. $x - 10 \geq -3$
3. $2x + 5 < 7$
4. $5x + 3 \leq -4$
5. $-3x \leq 9$
6. $4x \geq -32$
7. $\frac{x}{3} < 5$
8. $\frac{x}{-4} < 2$
9. $\frac{x}{-3} > 5 - 9$
10. $5x + 1 \leq 11$
11. $-3x + 6 < 24$
12. $6x + 5 > -11$
13. $x + 5 \leq 4x - 4$

ENTER WORD PROBLEMS . . .

Word problems using inequalities inevitably come along. But, before you surrender to confusion, let's look at an example and the steps to take to solve it.

Jim has twice as many pairs of socks as Debbie, and together, they have at least 12 pairs. What is the fewest number of pairs Debbie can have?

Start by assigning values to the information in this word problem. There are Debbie's socks and Jim's socks. Let x equal the number of pairs Debbie has and let $2x$ equal the number of pairs Jim has. Because the total pairs is *at least* 12, $x + 2x \geq 12$. Combine like terms on the left side: $3x \geq 12$.

Now, divide both sides by 3: $\frac{3x}{3} \geq \frac{12}{3}$, so $x \geq 4$. The fewest number of pairs of socks Debbie can have is 4. Jim has at least 8 pairs of socks. That's it! You've solved the mysterious inequality word problem.

PRACTICE LAP

DIRECTIONS: Use scratch paper to solve the following problems. You can check your answers at the end of this chapter.

14. You are treating a friend to a movie. You will buy 2 tickets and spend $8 on concessions. If you don't want to spend more than $20, how much can you spend on the tickets?

15. You are pricing lawn furniture and plan to buy 4 chairs. You don't want to spend more than $120. What is the most you can spend on 1 chair?

16. You are going to a restaurant for lunch. You have $15 to spend. Your beverage is $2.50, and you will leave a $2 tip. How much can you spend on the entrée?

17. There are only 2 hours left in the day and Amy still has 5 people left to interview. What is the maximum amount of time she can spend on each interview if she spends the same exact amount of time with each person?

SOLVING COMPOUND INEQUALITIES WITH ONE VARIABLE

A **compound inequality** is a combination of two or more inequalities. For example, take the compound inequality $-3 < x + 1 < 4$.

To solve this compound inequality, subtract 1 from all parts of the inequality: $-3 - 1 < x + 1 - 1 < 4 - 1$. Simplify: $-4 < x < 3$.

You can see that the solution set is all numbers between -4 and 3. On a graph, the solution set is

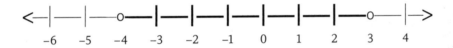

Here is another common type of a compound inequality. What would the solution set of $x < 4$ or $x > 7$ look like? Treat this the same way you handled the last example:

FUEL FOR THOUGHT

THE SOLUTION OF a compound inequality will often be one of two general cases. One is between two values, as shown in the following graph:

These inequalities will sometimes use the word *and*.

The other case is *less than or equal to* the smaller value or *greater than or equal to* the larger value, as in the following graph.

These inequalities will often use the word *or*.

A compound inequality usually involves two inequalities at one time. Solve them one at a time to make it easier.

PRACTICE LAP

DIRECTIONS: Graph the following inequalities on a number line. You can check your answers at the end of this chapter.

18. $x < -3$

19. $x > 1\frac{1}{2}$

20. $x \le 4$

21. $x \ge 0$

GRAPHING LINEAR INEQUALITIES OF TWO VARIABLES

Graphing linear inequalities of two variables has basically the same steps as graphing linear equations. Put the inequality in slope-intercept form and identify the slope and the *y*-intercept. When you are ready to graph, however, there are two differences.

When you are graphing a linear inequality using the symbol < or >, the line drawn is dashed. Points on the line *are not* part of the solution set. Inequalities with ≤ or ≥ are solid lines. Points on those lines *are* part of the solution set.

The other difference is that one side of the line is shaded to show all points in the plane that satisfy the inequality. Keep in mind that most inequalities have many, even an infinite number of solutions. Shading one side of a linear inequality is called shading the **half-plane**.

To find the side of the line that should be shaded, use a test point. If the test point makes the inequality true, then shade the side of the line where the test point is located. If the test point makes the inequality false, shade the opposite side of the line.

INSIDE TRACK

A GOOD TEST point to use is (0,0) because it is easy to substitute in and evaluate. The only time a test point of (0,0) should not be used is if it is located exactly on the line. In that case, you should choose a different test point.

Let's take a look at an example of graphing a linear inequality. Graph the inequality $y < -2x + 1$.

This inequality is in slope-intercept form; the slope is -2, and the *y*-intercept is 1. Before you begin graphing, look at the inequality symbol. Because the symbol is *less than*, the line should be a dashed line. Draw the line on a graph as you would draw a linear equation.

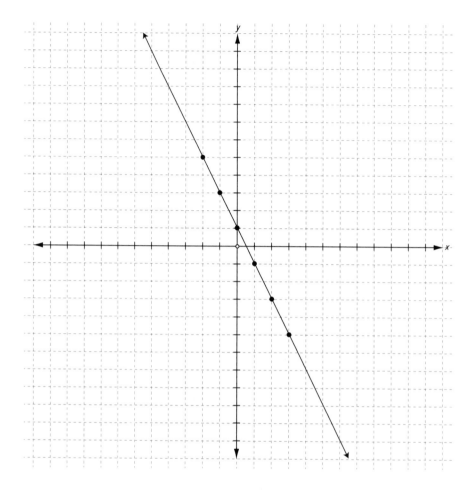

Now, use a test point to determine which side of the line should be shaded. Because (0,0) is not located on the line, you can use that point: $y <$ $-2x + 1$ becomes $0 < -2(0) + 1$. This simplifies to $0 < 0 + 1$, which is equal to $0 < 1$. Because this is a true statement, the point (0,0) is in the solution set of the inequality. On the graph, shade the side of the line that contains the origin. This is the side below the line. *All* points on this side of the inequality are solutions.

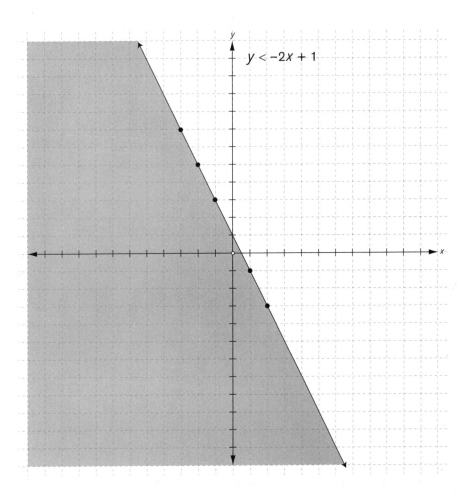

$y < -2x + 1$

In general, for inequalities using < and ≤, shade below the line. For inequalities using > and ≥, shade above the line.

PRACTICE LAP

DIRECTIONS: Graph the following inequalities on graph paper. You can check your answers at the end of this chapter.

22. $x + y > 1$

23. $6x + 2y \leq 8$

24. $-4x + 2y \geq 6$

25. $x - y > 5$

26. $3x - y < 2$

SPECIAL CASES OF LINEAR INEQUALITIES

There are two special cases of linear inequalities—one has a horizontal boundary line and the other has a vertical boundary line.

Inequalities with horizontal boundary lines have a slope of zero and are in the form $y < k$, $y \le k$, $y > k$, or $y \ge k$, where k is any number. The graph of the inequality $y < 2$ is

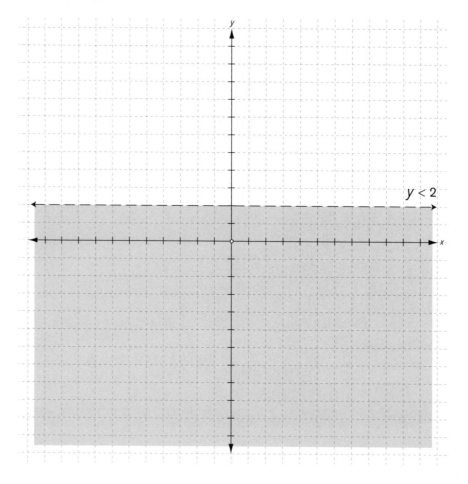

The inequality $y < 2$ is the same as the inequality $y < 0x + 2$. It has a slope of zero and a y-intercept of 2.

The graph of the inequality $y > -6$, which also has a horizontal boundary line, is

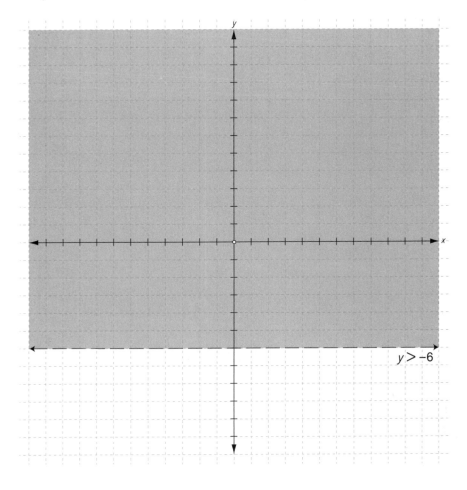

Inequalities with a vertical boundary line have undefined slope and are in the form $x < k$, $x \leq k$, $x > k$, or $x \geq k$, where k is any number. The graph of the inequality $x < 3$ is

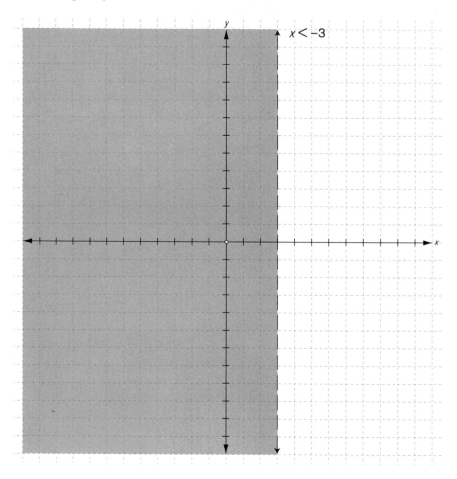

The graph of the inequality $x > -2$ is

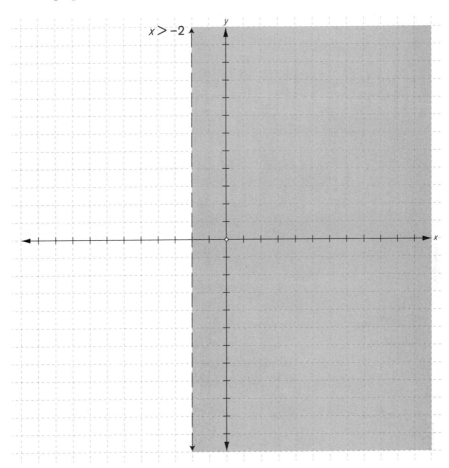

SOLVING SYSTEMS OF INEQUALITIES WITH TWO VARIABLES

Solving a system of inequalities is similar to solving a system of equations. The difference here is that you are looking for an overlapping in the shaded portions of the inequalities, not just one point of intersection.

Example

Graph the solution of the system of linear inequalities.

$y + x > 3$
$2y \leq 4x + 2$

In order to graph the solution to the system, graph both inequalities on the same set of axes and look for the region on the graph where the shaded sections overlap.

In the first inequality, subtract x from both sides to get the inequality in slope-intercept form. This inequality becomes $y > -x + 3$ with a slope of -1 and a y-intercept of 3. Graph this inequality on a set of axes making the line dashed. Shade above the line because the inequality has a $>$ symbol.

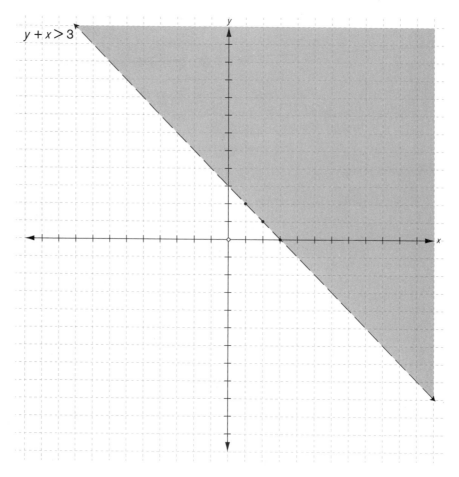

In the second inequality, divide both sides by 2 to get it into slope-intercept form. The result is the inequality $y \leq 2x + 1$ with a slope of 2 and a y-intercept of 1. Draw a solid line and shade the region below the line.

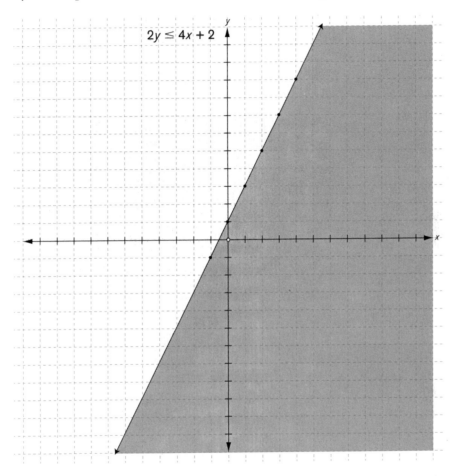

Now graph both inequalities on the same set of axes.

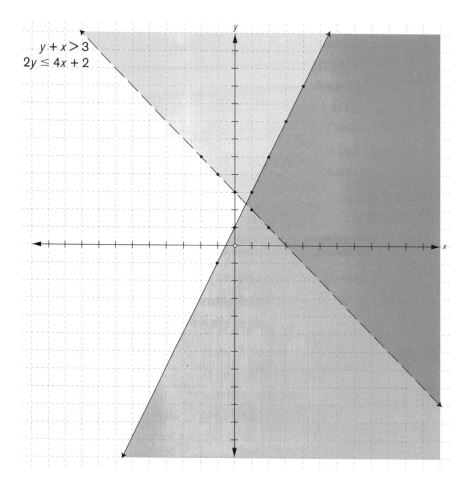

$y + x > 3$
$2y \leq 4x + 2$

The region where the shaded areas overlap is indicated by the darker shading in the figure. Any ordered pair in this region is a solution to the system of inequalities.

PACE YOURSELF

INEQUALITIES CAN BE graphed on a graphing calculator using the **Y =** screen. Try one! Type your inequality in as you would an equation and move your cursor to the far left of the equal sign. Use the Enter key to move through the choices. Choose the symbol for shading up for *greater than* inequalities and choose the symbol for shading down for *less than* inequalities.

A special type of system of inequalities occurs if the lines are parallel. Take, for instance, the system of inequalities $y > x + 3$ and $y < x - 1$. For the first inequality, graph a dashed line with slope of 1 and y-intercept of 3 and shade above the line. For the second inequality, graph a dashed line with slope of 1 and y-intercept of -1 and shade below the line.

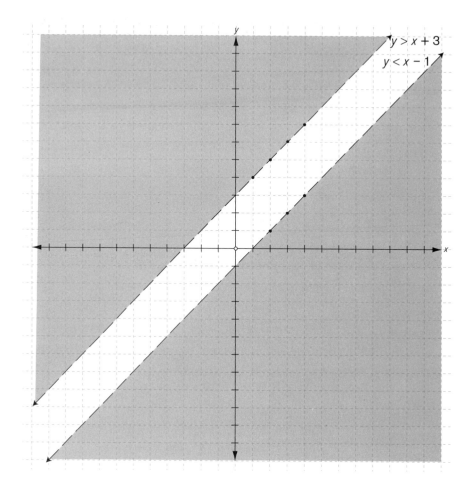

Notice that, on the graphs, there is not an intersection of the shaded areas. Therefore, there is no solution to this system of inequalities.

INSIDE TRACK

USE THE PROCESS of elimination to answer inequality multiple-choice questions. Look for traits such as open circles or a dashed line on a graph to match a given inequality.

PRACTICE LAP

DIRECTIONS: Solve the systems of inequalities graphically. You can check your answers at the end of this chapter.

27. $y > 4$
   $y < x + 2$

28. $y \geq 5$
   $x \leq 2$

29. $y < x + 2$
   $y < -x + 4$

ANSWERS

1. $x < 7$

2. $x \geq 7$

3. $x < 1$

4. $x \leq \frac{-7}{5}$ or $x \leq -1\frac{2}{5}$

5. $x \geq -3$

6. $x \geq -8$

7. $x < 15$

8. $x > -8$

9. $x < 12$

10. $x \leq 2$

11. $x > -6$

12. $x > \frac{-16}{6}$ (Reduce all fractions.)
   $x > \frac{-8}{3}$ (Another possible answer
   is $x > -2\frac{2}{3}$.)

13. $x \geq 3$

14. $2x + 8 \leq \$20, x \leq 6$

15. $4x \leq 120, x \leq \$30$; the most you
   can spend on a chair is $30.

16. $x + \$2.50 + \$2 \leq \$15, x \leq \$10.50$

17. $5x \leq 120, x \leq 24$ minutes,
   or $x \leq \frac{2}{5}$ hour

18.

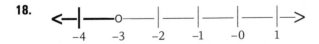

19.

20.

21.

22.

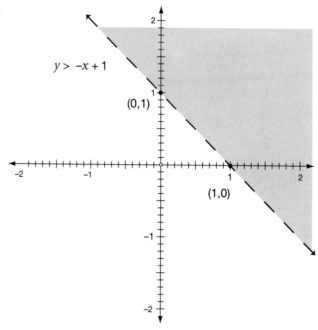

$y > -x + 1$

$(0,1)$

$(1,0)$

23.

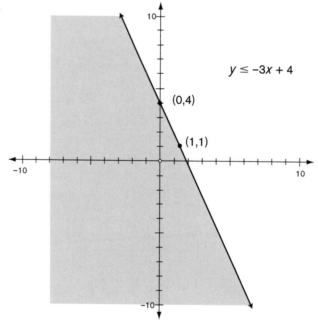

$y \leq -3x + 4$

$(0,4)$

$(1,1)$

24.

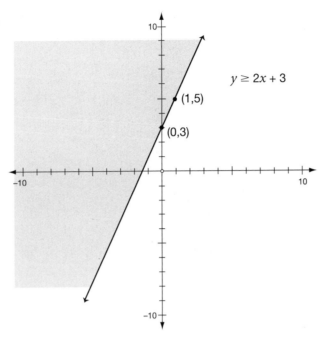

$y \geq 2x + 3$

(1,5)

(0,3)

25.

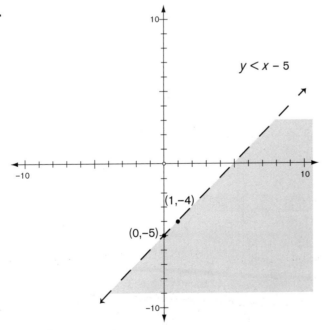

$y < x - 5$

(1,-4)

(0,-5)

26.

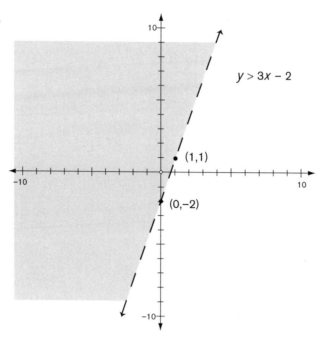

$y > 3x - 2$

(1,1)

(0,−2)

27.

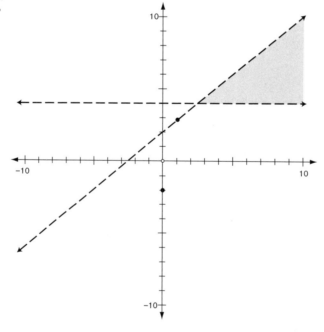

28.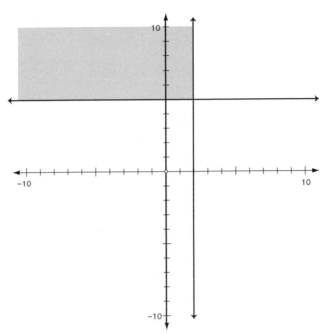

29.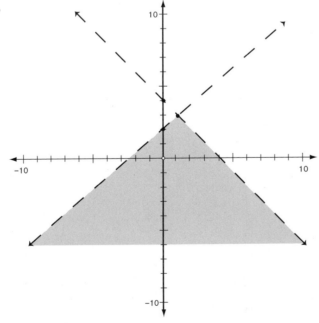

Puzzling Polynomials, Factoring, and Quadratic Equations

WHAT'S AROUND THE BEND

➥ Degree of Polynomials
➥ Combining Like Terms
➥ Multiplying Polynomials
➥ Factoring Polynomials
➥ Solving Quadratic Equations

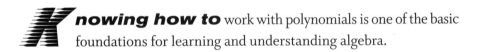 **nowing how to** work with polynomials is one of the basic foundations for learning and understanding algebra.

DEGREE OF POLYNOMIALS

Take a look at the following terms:

$5x$	Five is the coefficient, x is the variable, 1 is the exponent.
ab	One is the coefficient, both a and b are the variables, and both have an exponent of 1.
$-4x^3y$	Negative four is the coefficient, x and y are the variables, 3 is the exponent on x, and 1 is the exponent on y.

Because different terms of polynomials are separated by addition and subtraction, each of these examples represents a **monomial**. Even though there may be more than one number and/or variable involved, they are still monomials. Another important aspect of monomials is the degree of the monomial.

The **degree of a monomial** is the sum of the exponents of the variables. The degree of the term $3x$ is 1 because $3x = 3x^1$. The degree of the term x^2y^3 is 5 because 2 plus 3 equals 5.

COMBINING LIKE TERMS

You know that **like terms** are expressions that have exactly the same variable(s) and exponents and can be combined easily by adding or subtracting the coefficients.

$4x + 5x$	These are *like terms*, and the sum is $9x$.
$6x^2y + -11x^2y$	These are also *like terms*, and the sum is $-5x^2y$.
$3a - (5a)$	These are *like terms*, and the difference is $-2a$.
$9xy^2 + 9x^2y$	These are NOT *like terms* because the exponents of the variables are not exactly the same. They cannot be combined.

When you are subtracting polynomials in particular, it is imperative that you use the distributive property on each term in the polynomial being subtracted. Take the expression $(3x^2 - 4x - 5) - (2x^2 - 7x + 9)$. The subtraction sign in front of the polynomial $(2x^2 - 7x + 9)$ is treated as a -1. To eliminate the parentheses, each term individually will be changed to its inverse as the subtraction is distributed to the $2x^2$, $-7x$, and 9. The entire expression then becomes $3x^2 - 4x - 5 - 2x^2 + 7x - 9$. Complete the problem by combining *like terms*. Therefore, the simplified expression is $x^2 + 3x - 14$. When unlike terms cannot be combined, they create an expression called a **polynomial**.

FUEL FOR THOUGHT

A POLYNOMIAL IS the sum or difference of two or more monomials. Terms of polynomials are separated by addition and subtraction.

Some polynomials have specific names:

$3x^2$ is a *monomial* because there is one term.

$7x + 8y$ is a *binomial* because there are two terms.

$2x^2 + 6x - 7$ is a *trinomial* because there are three terms.

You have already learned how to find the degree of a monomial. You can also find the degree of different types of polynomials.

The **degree of a polynomial** is equal to the degree of the term of the polynomial with the *greatest* degree. The degree of the polynomial $3x^2 + 5x - 9$ is 2. The first term, $3x^2$, has degree 2 because the exponent on the variable is 2. This is the term with the greatest degree of this polynomial. The degree of the polynomial $5x^2y + 3xy - 2y$ is 3. The leading term of $5x^2y$ has an exponent of 2 on x and 1 on y: $2 + 1 = 3$. The second term has only degree 2 and the third term has only degree 1.

MULTIPLYING POLYNOMIALS

When multiplying by a monomial, use the distributive property to simplify if there is more than one term to be multiplied. Multiply coefficients by coefficients, and add the exponents of any like bases.

Examples

Multiply each of the following:

$(6x^3)(5xy^2) = 30x^4y^2$ Remember that $x = x^1$.

$2x(x^2 - 3) = 2x^3 - 6x$ Use the distributive property.

$x^3(3x^2 + 4x - 2) = 3x^5 + 4x^4 - 2x^3$ Use the distributive property.

When multiplying two binomials together, use an acronym called **FOIL** to help you remember how to multiply two binomials using the distributive property.

    F   Multiply the **first** terms in each set of parentheses.

    O   Multiply the **outer** terms in the parentheses.

    I   Multiply the **inner** terms in the parentheses.

    L   Multiply the **last** terms in the parentheses.

Example

$(y - 3)^2 =$

Set up the binomials:	$(y - 3)(y - 3) =$
Use the FOIL method to multiply the terms:	$y^2 - 3y - 3y + 3^2 =$
Combine like terms:	$y^2 - 6y + 3^2 =$
Find the square of 3:	$y^2 - 6y + 9$

This example is common on many standardized tests. It is an example of the square of a binomial difference. Another common situation is the square of a binomial sum.

INSIDE TRACK

WHEN FINDING THE square of a binomial difference, use the formula $(x - y)^2 = x^2 - 2xy + y^2$.

When finding the square of a binomial sum, use the formula $(x + y)^2 = x^2 + 2xy + y^2$.

In addition to the previous example, another common type of binomial multiplication problem involves finding the product of the sum and the difference of the same two values. The general form is $(x - y)(x + y)$ and will multiply to the binomial $x^2 - y^2$.

When multiplying any type of polynomial by another polynomial, you should always use the distributive property. Take, for example, the binomial $x + 2$ and the trinomial $2x^2 - 3x - 5$. In order to multiply these together, you must multiply the x from the binomial by each term of the trinomial. Then, repeat the process with the 2 from the binomial. This is what the process would look like. To multiply $(x + 2)(2x^2 - 3x - 5)$, multiply each term of the trinomial first by x and then 2: $2x^2 \cdot x - 3x \cdot x - 5 \cdot x + 2x^2 \cdot 2 - 3x \cdot 2 - 5 \cdot 2$. Now, simplify each term: $2x^3 - 3x^2 - 5x + 4x^2 - 6x - 10$. Combine *like terms* to get the simplified expression: $2x^3 + x^2 - 11x - 10$.

DIRECTIONS: Use scratch paper to multiply the following polynomials. You can check your answers at the end of this chapter.

1. $5(x - y + 2)$
2. $7x(x - 3)$
3. $8x^3(3x^2 + 2x - 5)$
4. $-6(x - y - 7)$
5. $3b(x^2 + 2xy + y)$
6. $(x + 2)(x + 4)$
7. $(x + 6)(x - 3)$
8. $(x - 6)(x - 2)$
9. $(x - 1)(x + 10)$
10. $(x + 1)(x + 1)$
11. $(x - 2)(x^2 - 2x + 1)$
12. $(2x + 3)(x^2 + 2x + 5)$

REVIEWING FACTORS AND GCF

The factors of a number or expression are the numbers and/or variables that can be multiplied to equal it. For example, the number 8 can be written as

$$1 \cdot 8 = 8 \text{ or } 2 \cdot 4 = 8$$

The numbers 1, 2, 4, and 8 are the factors of 8.

The expression $4x^2$ is the result of multiplying $4 \cdot x \cdot x$, so these are the factors of this expression.

When factoring, you will often look for the greatest common factor of the terms in the expression. The **greatest common factor**, or **GCF**, is the greatest quantity that is common between the terms.

Example

Find the GCF of 24 and 36.

The factors of 24 are 1, 2, 3, 4, 6, 8, 12, and 24.
The factors of 36 are 1, 2, 3, 4, 6, 9, 12, 18, and 36.
Although the numbers 1, 2, 3, 4, 6, and 12 appear in both lists, the greatest of these is 12. Therefore, 12 is the GCF of 24 and 36.

To find the GCF of expressions that involve variables, look for the common numerical factors and the number of common variables. For the terms $4x^2y$ and $6xy^2$, the common numerical factor is 2. For the variables, the first term has two x's and one y. The second term has one x and two y's. Because the GCF contains only the amount of factors that are common to both terms, the common variables are xy. Therefore, the GCF of $4x^2y$ and $6xy^2$ is $2xy$.

FACTORING POLYNOMIALS

Factoring polynomials is the reverse of multiplying them together. The process of factoring polynomials usually falls into one of these categories: factoring out the GCF, finding the difference between two perfect squares, factoring a trinomial, and factoring by grouping. The goal of factoring is to factor as much as you can. In other words, factor *completely*. Here are some examples of each type.

Factoring out the GCF

In this type of problem, you will see that each term has something in common. This may be a numerical factor, variable factor, or a combination of both.

Let's work with $2x^3 + 2 = 2(x^3 + 1)$. Put in front of the parentheses the common factor of 2. The terms remaining within parentheses are the original terms divided by 2. To check that the factoring is correct, multiply the factored answer using the distributive property:

$$2(x^3 + 1) = 2 \cdot x^3 + 2 \cdot 1 = 2x^3 + 2$$

Factoring the Difference between Two Perfect Squares

This type of situation will present two perfect square terms separated by subtraction. Numerical perfect squares are the result of multiplying two of the same integers together like $5 \cdot 5 = 25$ or $9 \cdot 9 = 81$. Variables that are perfect squares will have an exponent that is an even number. For example, because $x \cdot x = x^2$, then $\sqrt{x^2} = x$. In order to factor the difference between two squares, take the square root of each perfect square and express the difference between them as one factor and their sum as the other.

For example, $x^2 - 9 = (x - 3)(x + 3)$ because $\sqrt{x^2} = x$ and $\sqrt{9} = 3$.

To check to make sure these factors are correct, multiply using **FOIL** (the distributive property).

Likewise, $4c^2 - 49 = (2c - 7)(2c + 7)$ because $\sqrt{4c^2} = 2c$ and $\sqrt{49} = 7$.

The difference of two squares is factored $x^2 - y^2 = (x - y)(x + y)$. The sum of two squares is **not** factorable.

Factor the Trinomial in the Form $ax^2 + bx + c$

In order to factor a binomial in this form, use backward **FOIL**. For situations in which the value of $a = 1$, the factors will contain two numbers whose sum is b and whose product is c. Take the following examples.

$$x^2 + 5x + 6$$

Because $x^2 = x \cdot x$, the leading terms in the parentheses will be both x's. In this problem, $a = 1$, so find two numbers that add to **b**, which is 5 and multiply to equal **c**, which in this case is 6. These two numbers are 2 and 3. This makes the factors $(x + 2)(x + 3)$. To check to make sure these factors are correct, multiply them using **FOIL** (distributive property).

$$x^2 - 3x - 10$$

In this case, you are looking for two numbers whose sum is -3 and whose product is -10. These two numbers are 2 and -5. Therefore, the factors are $(x + 2)(x - 5)$.

FUEL FOR THOUGHT

TO FACTOR A quadratic equation in the form $x^2 + bx + c = 0$ (the value of $a = 1$), the factors will be in the form $(x + \_)(x + \_)$ where the numerical factors have a **sum** of b and a **product** of c.

$x^2 - 9x + 18$

This trinomial has a b value of -9 and a c value of 18. Two numbers that add to -9 and whose product is 18 are -3 and -6. The factors are $(x - 3)(x - 6)$.

When factoring a trinomial in this form where the value of a is not 1, you may have to do a little more work. Very often, using trial and error with **FOIL** can lead you to the correct factors. Here's an example.

$2x^2 + 5x - 3$

Factor by performing **FOIL** backward, but keep in mind that because $a = 2$, the outside terms when multiplied will be affected. First, find factors of $2x^2$. These will be $2x$ and x and will be the leading terms in the parentheses. Now you need factors of -3 that will combine with $2x$ and x to make a middle term of $5x$. Find the factors by trial and error.

Begin by trying -3 and 1 as the factors of -3. The factors would be $(2x - 3)(x + 1)$. When multiplied together, they equal $2x^2 + 2x - 3x - 3$. Combine like terms to get $2x^2 - 1x - 3$, which is equal to $2x^2 - x - 3$. Because this is not the original trinomial, these are not the correct factors.

Switch the -3 and 1 to the opposite binomials. The factors would be $(2x + 1)(x - 3)$. When multiplied together, they equal $2x^2 - 6x + 1x - 3$. Combine like terms to get $2x^2 - 5x - 3$. Because this is not the original trinomial, these are not the correct factors.

Switch the signs on the numbers to make the factors of -3 to be 3 and -1. The factors of the trinomial would then be $(2x + 3)(x - 1)$. When multiplied together, they equal $2x^2 - 2x + 3x - 3$. Combine like terms to get $2x^2 + 1x - 3$, which is equal to $2x^2 + x - 3$. Because this is not the original trinomial, these are not the correct factors.

Switch the –1 and 3 to the opposite binomials. The factors become $(2x - 1)$ $(x + 3)$. When multiplied together, they equal $2x^2 + 6x - 1x - 3$. Combine like terms: $2x^2 + 5x - 3$. Because this is the original trinomial, these are the correct factors.

PRACTICE LAP

DIRECTIONS: Use scratch paper to factor the following polynomials. You can check your answers at the end of this chapter.

13. $3x + 12$

14. $11a + 33b$

15. $3a + 6b + 12c$

16. $2x^2 + 3x$

17. $5x + 9$

18. $16x^2 + 20x$

19. $x^2y + 3x$

20. $8x^3 - 2x^2 + 4x$

21. $a^2 - 49$

22. $b^2 - 121$

23. $4x^2 - 9$

24. $x^2 + y^2$

25. $r^2 - s^2$

26. $36b^2 - 100$

27. $a^6 - b^6$

28. $x^2 + 4x + 4$

29. $x^2 + 6x + 8$

30. $x^2 - 4x + 3$

31. $x^2 + 7x + 12$

32. $x^2 - 10x + 16$

33. $2x^2 + 7x + 3$

34. $5x^2 + 13x - 6$

35. $7x^2 + 3x - 4$

36. $4x^2 - 21x + 5$

37. $6x^2 + x - 7$

FACTORING BY GROUPING

Sometimes, you may encounter a polynomial with four terms. The strategies discussed previously may not work with this situation. Another factoring technique to try in this case is factoring by grouping. Here, you will pair together terms with a common factor. After factoring out that common factor from each pair, the terms left in the parentheses are often the same. These terms can then be factored out to create the factored polynomial. Note the following example of factoring by grouping.

Take the polynomial $ax + bx - 4a - 4b$. The first pair of terms has a common factor of x, and the second pair of terms has a common factor of -4. Factor each pair separately. The resulting expression is $x(a + b) - 4(a + b)$. Now, each of these terms has a factor of $(a + b)$. Factoring $a + b$ out in front results in the expression $(a + b)(x - 4)$, which gives the factors of the polynomial.

Factoring by grouping is usually the most helpful in expressions that have more than three terms.

FACTORING COMPLETELY

Many times, you will be presented with a polynomial whereby the factoring takes place over more than one step. It is very important that polynomial expressions are factored completely. Here are some examples where the factoring takes place over more than one step.

$2x^2 - 128$

First, factor a 2 out of both terms. The expression then becomes $2(x^2 - 64)$. Located in parentheses is the difference between two squares, which is also factorable. The final factors are $2(x - 8)(x + 8)$.

FUEL FOR THOUGHT

WHEN FACTORING POLYNOMIALS, use the following steps:

1. Always look for a greatest common factor of each term first.

2. Look for the difference between two perfect squares.

3. Factor the trinomial performing **FOIL** backward.

4. If there are four or more terms, try factoring by grouping.

What about the polynomial $d^4 - 16$? Because these terms do not have a common factor, start by factoring the difference between two squares. $(d^2 - 4)(d^2 + 4)$. Of these two factors, the first binomial is also the difference between two squares and becomes $(d - 2)(d + 2)$. The second binomial is the sum of two squares and is not factorable. Therefore, the factors are $(d - 2)(d + 2)(d^2 + 4)$.

Try one more: $2n^3 - 14n^2 + 24n$. First, factor out the common factor of $2n$ from each term. This leaves you with $2n(n^2 - 7n + 12)$. The trinomial in the parentheses can be factored. Two numbers that add to -7 and whose product is 12 are -3 and -4, so the factors are $(n - 3)(n - 4)$. Therefore, the factors for the entire polynomial are $2n(n - 3)(n - 4)$.

SOLVING QUADRATIC EQUATIONS

An equation in the form $0 = ax^2 + bx + c$, where a, b, and c are real numbers, is a **quadratic equation**. Quadratic equations contain a term in which the exponent on x is two—in other words, it contains an x^2. Two ways to solve quadratic equations algebraically are factoring, if it is possible for that equation, or by using the quadratic formula.

FUEL FOR THOUGHT

THE ZERO PRODUCT property states that if an equation is set equal to zero, then at least one of the factors must equal zero. If $xy = 0$, then either $x = 0$ or $y = 0$.

By Factoring

In order to factor the quadratic equation, it first needs to be in standard form. This form is $0 = ax^2 + bx + c$. In most cases, the factors of the equations involve two numbers whose sum is b and product is c. Be sure to factor the quadratic completely. After factoring, set each factor equal to zero and solve for x. In this step, you are using the zero product property as mentioned in the sidebar. The resulting values are called the **roots** of the equation.

Try this method by solving the following for x: $x^2 - 100 = 0$.

This equation is already in standard form. This equation is a special case; it is the difference between two perfect squares. To factor this, find the square root of both terms. The square root of the first term x^2 is x. The square root of the second term 100 is 10. Then the two factors are $x - 10$ and $x + 10$. The equation $x^2 - 100 = 0$ then becomes $(x - \underline{10})(x + \underline{10}) = 0$. Set each factor equal to zero and solve: $x - 10 = 0$ or $x + 10 = 0$. So, $x = 10$ or $x = -10$. The solution is $\{10, -10\}$.

Look at $x^2 + 14x = -49$. This equation needs to be put into standard form by adding 49 to both sides of the equation. $x^2 + 14x + 49 = -49 + 49$, or $x^2 + 14x + 49 = 0$. The factors of this trinomial will be two numbers whose sum is 14 and whose product is 49. The factors are $x + 7$ and $x + 7$ because $7 + 7 = 14$ and $7 \cdot 7 = 49$. The equation becomes $(x + \underline{7})(x + \underline{7}) = 0$. Set each factor equal to zero and solve: $x + 7 = 0$ or $x + 7 = 0$. So, $x = -7$ or $x = -7$. Because both factors were the same, this was a perfect square trinomial. The solution is $\{-7\}$.

One more just to drive the point home. Solve $x^2 = 42 + x$. This equation needs to be put into standard form by subtracting 42 and x from both sides of the equation: $x^2 - x - 42 = 42 - 42 + x - x$, or $x^2 - x - 42 = 0$. Because the sum of 6 and -7 is -1, and their product is -42, the equation factors to $(x + \underline{6})(x - \underline{7}) = 0$. Set each factor equal to zero and solve: $x + 6 = 0$ or $x - 7 = 0$. So, $x = -6$ or $x = 7$. The solution is $\{-6, 7\}$.

By Quadratic Formula

If the equation is not factorable, or if you are having trouble finding the factors, use the quadratic formula. Solving by using the quadratic formula will work for any quadratic equation and is necessary for those that are not factorable. The quadratic formula is $x = \frac{-b \pm \sqrt{b^2 - 4ac}}{2a}$. To use this formula, first put the equation into standard form. Then, identify a, b, and c from the equation and substitute these values into the quadratic formula.

Let's test this. Solve $x^2 + 4x = 3$ for x. Put the equation in standard form: $x^2 + 4x - 3 = 0$. Because this equation is not factorable, use the quadratic formula by identifying the values of a, b, and c and then substituting them into the formula. For this particular equation, $a = 1$, $b = 4$, and $c = -3$.

$$x = \frac{-b \pm \sqrt{b^2 - 4ac}}{2a}$$

$$x = \frac{-4 \pm \sqrt{4^2 - 4(1)(-3)}}{2(1)}$$

$$x = \frac{-4 \pm \sqrt{16 + 12}}{2}$$

$$x = \frac{-4 \pm \sqrt{28}}{2}$$

$$x = \frac{-4}{2} \pm \frac{2\sqrt{7}}{2} \text{ (Remember that } \sqrt{28} \text{ reduces to } 2\sqrt{7}.)$$

$$x = -2 \pm \sqrt{7}$$

The solution is $\{-2 + \sqrt{7}, -2 - \sqrt{7}\}$. These roots are irrational because they contain a radical in the simplified solution. The square root of any non-perfect square is irrational.

INSIDE TRACK

IF A QUADRATIC equation is difficult to factor or is unfactorable, use the quadratic formula to find the roots: $x = \frac{-b \pm \sqrt{b^2 - 4ac}}{2a}$.

DIRECTIONS: Use scratch paper to solve the following problems. You can check your answers at the end of this chapter.

38. $x^2 - 36 = 0$

39. $2x^2 - 50 = 0$

40. $x^2 = 49$

41. $4x^2 = 100$

42. $x^2 - 50 = -1$

43. $x^2 + x - 2 = 0$

44. $x^2 + 7x - 18 = 0$

45. $x^2 + 4x = 45$

46. $x^2 - 6 = 30 - 3x^2$

47. $x^2 + 11x = -24$

ANSWERS

1. $5x - 5y + 10$

2. $7x^2 - 21x$

3. $24x^5 + 16x^4 - 40x^3$

4. $-6x + 6y + 42$

5. $3bx^2 + 6bxy + 3by$

6. $x^2 + 6x + 8$

7. $x^2 + 3x - 18$

8. $x^2 - 8x + 12$

9. $x^2 + 9x - 10$

10. $x^2 + 2x + 1$

11. $x^3 - 4x^2 + 5x - 2$

12. $2x^3 + 7x^2 + 16x + 15$

13. $3(x + 4)$

14. $11(a + 3b)$

15. $3(a + 2b + 4c)$

16. $x(2x + 3)$

17. prime—can't be factored

18. $4x(4x + 5)$

19. $x(xy + 3)$

20. $2x(4x^2 - x + 2)$

21. $(a + 7)(a - 7)$

22. $(b + 11)(b - 11)$

23. $(2x - 3)(2x + 3)$

24. prime—can't be factored

25. $(r + s)(r - s)$

26. $(6b + 10)(6b - 10)$

27. $(a^3 + b^3)(a^3 - b^3)$

28. $(x + 2)(x + 2)$

29. $(x + 4)(x + 2)$

30. $(x - 1)(x - 3)$

31. $(x + 4)(x + 3)$

32. $(x - 8)(x - 2)$

33. $(2x + 1)(x + 3)$

34. $(5x - 2)(x + 3)$

35. $(7x - 4)(x + 1)$

36. $(4x - 1)(x - 5)$

37. $(6x + 7)(x - 1)$

38. $6, -6$

39. $5, -5$

40. $7, -7$

41. $5, -5$

42. $7, -7$

43. $-2, 1$

44. $-9, 2$

45. $-9, 5$

46. $-3, 3$

47. $-8, -3$

Translating the Language of Algebra

WHAT'S AROUND THE BEND

- ➡ Translating Sentences into Expressions and Equations
- ➡ Solving Consecutive Integer Problems
- ➡ Mixture Problems
- ➡ Ratio Problems
- ➡ Geometry Problems
- ➡ Work Problems

Translating sentences and word problems into mathematical expressions and equations is similar to translating from one language into another. The key words are the vocabulary that tells what operations should be done and the order in which they should be evaluated.

CAUTION!

WHEN TRANSLATING KEY words, the phrases *less than* and *greater than* in an algebraic expression do not translate in the same order as they are written in the sentence. For example, when translating the expression *five less than 12*, the correct expression is 12 – 5, not 5 – 12.

Here is an example of a problem for which knowing the key words is necessary.

Twenty less than five times a number is equal to the product of ten and the number. What is the number?

Let's let *x* equal the number you are trying to find. Now, translate the sentence piece by piece, and then solve the equation.

Twenty less than five times a number equals the product of 10 and x.

$5x - 20 = 10x$

The equation is $5x - 20 = 10x$.

Subtract 5x from both sides: $5x - 5x - 20 = 10x - 5x$

Divide both sides by 5: $\frac{-20}{5} = \frac{5x}{5}$

The result: $-4 = x$

In this particular example, it is important to realize that the key words *less than* tell you to subtract from the number and the key word *product* reminds you to multiply.

PRACTICE LAP

DIRECTIONS: Use scratch paper to solve the following problems. You can check your answers at the end of this chapter.

1. What expression is equivalent to "the sum of three times a number and four"?

2. What expression is equivalent to "three less than five times a number"?

3. The product of −3 and a number is equal to −36. What is the number?

4. Twice a number increased by 10 is equal to 32 less than the number. Find the number.

PROBLEM SOLVING WITH WORD PROBLEMS

There are a variety of different types of word problems you will encounter on various standardized tests and other kinds of important tests. To help with these types of problems, always begin first by figuring out what you need to solve for and defining your variable(s) as what is unknown. Then, write and solve an equation that matches the question asked. The first type of problem is consecutive integer problems.

Defining Consecutive Integers

Consecutive integers are integers that differ by one, listed in numerical order. Examples of three consecutive integers are 3, 4, 5 or −11, −10, −9. Consecutive *even* integers are numbers like 10, 12, 14 or −22, −20, −18. Consecutive *odd* integers are numbers like 7, 9, 11. Consecutive even integers differ by 2, as do consecutive odd integers.

To solve a consecutive integer problem, first define the integers. Then, write an equation that matches the sentence in the problem.

Example

What is the greater of two consecutive even integers whose sum is 430?

Because these are two consecutive *even* integers, let $x =$ the lesser even integer and let $x + 2 =$ the greater even integer. Because *sum* is a key word for addition, the equation is $(x) + (x + 2) = 430$. Combine like terms on the left side of the equation: $2x + 2 = 430$. Subtract 2 from both sides of the equation: $2x + 2 - 2 = 430 - 2$. This simplifies to $2x = 428$. Divide each side by 2: $\frac{2x}{2} = \frac{428}{2}$. So, $x = 214$, which is the lesser integer. Thus, the greater even integer is $x + 2 = 216$.

PRACTICE LAP

DIRECTIONS: Use scratch paper to solve the following problems. You can check your answers at the end of this chapter.

5. The sum of two consecutive even integers is 94. What are the integers?

6. What is the lesser of two consecutive positive integers whose product is 240?

7. What is the greater of two consecutive negative integers whose product is 462?

Mixture Problems

Mixture questions will present you with two or more different types of "objects" to be mixed together. Some common types of mixture scenarios are combining different amounts of money at different interest rates, different amounts of solutions at different concentrations, and different amounts of food (candy, coffee, etc.) that have different prices per pound.

Example

How many pounds of coffee that costs $4 per pound need to be mixed with 10 pounds of coffee that costs $6.40 per pound to create a mixture of coffee that costs $5.50 per pound?

Remember that the total amount spent in each case will be the price per pound times how many pounds in the mixture. Therefore, if you let x = the number of pounds of $4 coffee, then $4(x)$ is the amount of money spent on $4 coffee, $6.40(10)$ is the amount spent on $6.40 coffee, and $5.50(x + 10)$ is the total amount spent. Write an equation that adds the first two amounts and sets it equal to the total amount: $4(x) + 6.40(10) = 5.50(x + 10)$

Multiply through the equation: $4x + 64 = 5.5x + 55$. Subtract $4x$ from both sides: $4x - 4x + 64 = 5.5x - 4x + 55$. Now, subtract 55 from both sides: $64 - 55 = 1.5x + 55 - 55$. Divide both sides by 1.5: $\frac{9}{1.5} = \frac{1.5x}{1.5}$. You can see that $6 = x$. You need 6 pounds of the $4 per pound coffee.

PRACTICE LAP –

DIRECTIONS: Use scratch paper to solve the following problems. You can check your answers at the end of this chapter.

8. Jake invested money in two different accounts, part at 12% interest and the rest at 15% interest. The amount invested at 15% was twice the amount at 12%. How much was invested at 12% if the total interest earned was $1,134?

9. Martha bought *x* pounds of coffee that cost $3 per pound and 15 pounds of coffee at $2.50 per pound for the company picnic. Find the total number of pounds of coffee purchased if the average cost per pound is $2.70.

10. The junior class of a high school bought two different types of candy to sell at a school fund-raiser. They purchased 50 pounds of candy at $2.25 per pound and *x* pounds at $1.90 per pound. What is the total number of pounds they bought if the total amount of money spent on candy was $169.50?

Ratio Problems

Questions of this type often seem difficult because they usually contain more variables than actual numbers. However, this type of question can be made simpler if numbers are substituted for the letters. Proportions can also be used to help in these situations, but be sure to use the same units for both ratios.

Example

If Jed earns a total of *a* dollars in *b* hours, how many dollars will he earn in *c* hours?

Think about the problem with actual numbers. If Jed makes 20 dollars in 2 hours, he makes 10 dollars per hour. This was calculated by dividing $20 by 2 hours—in other words, $a \div b$. To calculate how much he will make in *c* hours, take the amount he makes per hour and multiply by *c*. Therefore, the expression $(a \div b) \cdot c$ simplifies to $\frac{a}{b} \times c = \frac{ac}{b}$.

This problem can also be looked at using a proportion. Write the ratios lining up the labels and then cross multiply to solve for the unknown. For this problem, the proportion would be $\frac{a \text{ dollars}}{b \text{ hours}} = \frac{? \text{ dollars}}{c \text{ hours}}$. Cross multiply to get ($?$ dollars) $\cdot b = a \cdot c$. Divide by b to get ($?$ dollars) $= \frac{ac}{b}$.

PRACTICE LAP

DIRECTIONS: Use scratch paper to solve the following problems. You can check your answers at the end of this chapter.

11. If Tonya fills x containers in m minutes, how many minutes will it take to fill y containers?
12. If e eggs are needed to make c cookies, how many eggs are needed to make $10c$ cookies?
13. A car travels m miles in h hours. At that rate, how many miles does it travel in 90 minutes?

Geometry Problems

The key to solving many of the geometry problems presented on exams is knowing the correct formula to apply. Much of the time, the formula necessary is a well-known formula, such as area or perimeter.

FUEL FOR THOUGHT

THE FORMULA FOR the *area* of a parallelogram is *Area* = *base* × *height*. The family of parallelograms also includes squares, rhombuses, and rectangles. Area of any triangle is *Area* = $\frac{1}{2}$ × *base* × *height*, because a triangle is half of a parallelogram. Keep in mind that in this type of problem, *base* and *height* can be interchanged with *length* and *width*.

Perimeter is the distance around an object. Perimeter is found by adding the lengths of the sides of the object together.

The following example combines a commonly known formula with algebra to solve for the dimensions of a geometric shape.

Example

The length of a rectangle is two meters more than its width. The area of the rectangle is 48 square meters. What is the length of the rectangle?

Always start by defining your unknowns in the problem. Let w = the width of the rectangle. Because the length is two more than the width, let $w + 2 = l$. To find the area of the rectangle, use the formula *Area = length* × *width*, which for a rectangle is the same as *Area = base* × *height*. Because you know that the area is 48, the formula *Area = length* × *width* becomes $48 = w(w + 2)$. To solve, distribute the right side of the equal sign to get $48 = w^2 + 2w$. Subtract 48 from both sides to get the equation in standard form: $0 = w^2 + 2w - 48$. Factor the quadratic equation into two binomials: $0 = (w + 8)(w - 6)$. Set each factor equal to zero and solve for w: $w + 8 = 0$ or $w - 6 = 0$. Therefore, $w = -8$ or 6. Because you are finding the dimensions of a geometric figure, reject the negative value. The width of the rectangle is 6, and the length is $6 + 2 = 8$.

CAUTION!

BE SURE TO reject any negative value in a geometry problem when solving for a dimension of a figure. You cannot have a negative length, width, or distance.

---- PRACTICE LAP ----

DIRECTIONS: Use scratch paper to solve the following problems. You can check your answers at the end of this chapter.

14. The perimeter of a rectangle is 42 inches. What is the measure of its width if its length is 3 inches greater than its width?

15. Sally owns a rectangular field that has an area of 1,200 square feet. The length of the field is 2 more than twice the width. What is the width of the field?

16. The perimeter of a parallelogram is 100 centimeters. The length of the parallelogram is 5 centimeters more than the width. Find the length of the parallelogram.

Work Problems

Work problems often present the scenario of two people working to complete the same job. To solve this particular type of problem, think about how much of the job will be completed in 1 hour. For example, if someone can complete a job in 5 hours, then $\frac{1}{5}$ of the job is completed in 1 hour. If a person can complete a job in x hours, then $\frac{1}{x}$ of the job is completed in 1 hour.

Example

Jason can mow a lawn in 2 hours. Ciera can mow the same lawn in 4 hours. If they work together, how many hours will it take them to mow the same lawn?

Think about how much of the lawn each person completes individually. Because Jason can finish in 2 hours, in 1 hour he completes $\frac{1}{2}$ of the lawn. Because Ciera can finish in 4 hours, then in 1 hour she completes $\frac{1}{4}$ of the lawn. If you let x equal the time it takes both Jason and Ciera working together, then $\frac{1}{x}$ is the amount of the lawn they finish in 1 hour working together. Then use the equation $\frac{1}{2} + \frac{1}{4} = \frac{1}{x}$ and solve for x. Multiply each term by the LCD of $4x$: $4x(\frac{1}{2}) + 4x(\frac{1}{4}) = 4x(\frac{1}{x})$. The equation becomes $2x + x = 4$. Combine like terms: $3x = 4$. Now, divide each side by 3: $\frac{3x}{3} = \frac{4}{3}$. You can see that $x = 1\frac{1}{3}$ hours. Because $\frac{1}{3}$ of 1 hour is $\frac{1}{3}$ of 60 minutes, which is 20 minutes, the correct answer is 1 hour 20 minutes.

PRACTICE LAP

DIRECTIONS: Use scratch paper to solve the following problems. You can check your answers at the end of this chapter.

17. Heather can remodel a kitchen in 10 hours; Steve can do the same job in 15 hours. If they work together, how many hours will it take them to redo the kitchen?

18. Anthony can seal the driveway in 150 minutes; John can seal the same driveway in 120 minutes. How many hours will it take them to seal the driveway if they do it together?

ANSWERS

1. *Sum* is a key word for addition, and the translation of "three times a number" is $3x$. This is telling you to add the two parts of the sentence, $3x$ and 4. The complete translation would be $3x + 4$.

2. When the key words *less than* appear in the sentence, you know that you will subtract 3 from the next part of the sentence, so it will appear at the end of the expression. "Five times a number" can be represented by $5x$. The correct translation is $5x - 3$.

3. Let x equal a number. Because *product* is a key word for multiplication, the equation is $-3 \cdot x = -36$. Divide both sides by -3: $\frac{-3x}{-3} = \frac{-36}{-3}$. The variable is now alone: $x = 12$.

4. Let x equal a number. Now translate each part of the sentence.

Twice a number increased by 10 is	$2x + 10$
Thirty-two less than a number is	$x - 32$
Set them equal as they are in the sentence:	$2x + 10 = x - 32$
Subtract x from both sides of the equation:	$2x - x + 10 = x - x - 32$
Simplify:	$x + 10 = -32$
Subtract 10 from both sides of the equation:	$x + 10 - 10 = -32 - 10$
The variable is now alone:	$x = -42$

5. Two consecutive even integers are numbers in order, like 4, 6 or −30, −32, each of which is two integers apart. Let x equal the first consecutive even integer. Let $x + 2$ equal the second consecutive even integer. *Sum* is a key word for addition, so the equation becomes $(x) + (x + 2) = 94$. Combine like terms on the left side of the equation: $2x + 2 = 94$. Subtract 2 from both sides of the equation: $2x + 2 - 2 = 94 - 2$. Simplify: $2x = 92$. Divide each side of the equation by 2: $\frac{2x}{2} = \frac{92}{2}$. $x = 46$. Therefore, $x + 2 = 48$. The integers are 46 and 48.

6. Let x equal the lesser integer and let $x + 1$ equal the greater integer. Because *product* is a key word for multiplication, the equation is $x(x + 1) = 240$. Multiply using the distributive property on the left side of the equation: $x^2 + x = 240$. Put the equation in standard form and set it equal to zero: $x^2 + x - 240 = 0$. Factor the trinomial: $(x - 15)(x + 16) = 0$. Set each factor equal to zero and solve: $x - 15 = 0$ or $x + 16 = 0$. So, $x = 15$ or $x = -16$. Because you are looking for a positive integer, reject the x-value of −16. If $x = 15$, then $x + 1 = 16$. Therefore, the lesser positive integer would be 15.

7. Let x equal the lesser integer and let $x + 1$ equal the greater integer. Because *product* is a key word for multiplication, the equation is $x(x + 1) = 462$. Multiply using the distributive property on the left side of the equation: $x^2 + x = 462$. Put the equation in standard form and set it equal to zero: $x^2 + x - 462 = 0$. Factor the trinomial: $(x - 21)(x + 22) = 0$. Set each factor equal to zero and solve: $x - 21 = 0$ or $x + 22 = 0$. So, $x = 21$ or $x = -22$. Because you are looking for a negative integer, reject the x-value of 21. Therefore, $x = -22$ and $x + 1 = -21$. The greater negative integer is −21.

8. Let x equal the amount invested at 12% interest. Let y equal the amount invested at 15% interest. Because the amount invested at 15% is twice the amount at 12%, then $y = 2x$. Because the total interest was $1,134, use the equation $0.12x + 0.15y = 1,134$. You have two equations with two variables. Use the second equation $0.12x + 0.15y = 1,134$ and substitute $(2x)$ for y: $0.12x + 0.15(2x) = 1,134$. Multiply on the left side of the equal sign: $0.12x + 0.3x = 1,134$. Combine like terms: $0.42x = 1,134$. Divide both sides by 0.42: $\frac{0.42x}{0.42} = \frac{1,134}{0.42}$. Therefore, $x = \$2,700$, which is the amount invested at 12% interest.

9. Let x equal the amount of coffee at $3 per pound. Let y equal the total amount of coffee purchased. If there are 15 pounds of coffee at $2.50 per pound, then the total amount of coffee can be expressed as $y = x + 15$. Use the equation $3x + 2.50(15) = 2.70y$, because the total amount of pounds y costs $2.70 per pound. To solve, substitute $y = x + 15$ into $3x + 2.50(15) = 2.70y$. So, $3x + 2.50(15) = 2.70(x + 15)$. Multiply on the left side and use the distributive property on the right side: $3x + 37.50 = 2.70x + 40.50$. Subtract $2.70x$ on both sides: $3x - 2.70x + 37.50 = 2.70x - 2.70x + 40.50$. Subtract 37.50 from both sides: $0.30x + 37.50 - 37.50 = 40.50 - 37.50$. Divide both sides by 0.30: $\frac{0.30x}{0.30} = \frac{3}{0.30}$. So, $x = 10$ pounds, which is the amount of coffee that costs $3 per pound. Therefore, the total amount of coffee is $10 + 15$, which is 25 pounds.

10. Let x equal the amount of candy at $1.90 per pound. Let y equal the total number of pounds of candy purchased. If there are also 50 pounds of candy at $2.25 per pound, then the total amount of candy can be expressed as $y = x + 50$. Use the equation $1.90x + 2.25(50) = \$169.50$, because the total amount of money spent was $169.50. Multiply on the left side: $1.90x + 112.50 = 169.50$. Subtract 112.50 from both sides: $1.90x + 112.50 - 112.50 = 169.50 - 112.50$. Divide both sides by 1.90: $\frac{1.90x}{1.90} = \frac{57}{1.90}$. So, $x = 30$ pounds, which is the amount of candy that costs $1.90 per pound. Therefore, the total amount of candy is $30 + 50$, which is 80 pounds.

11. Write the ratios as a proportion lining up the word labels to help: $\frac{x \text{ containers}}{m \text{ minutes}} = \frac{y \text{ containers}}{? \text{ minutes}}$. Cross multiply and solve for the unknown number of minutes: $(? \text{ minutes}) \cdot x = m \cdot y$. Divide both sides by x to get the solution: $(? \text{ minutes}) = \frac{my}{x}$.

12. Write the ratios as a proportion lining up the word labels to help: $\frac{e \text{ eggs}}{c \text{ cookies}} = \frac{? \text{ eggs}}{10c \text{ cookies}}$. Cross multiply and solve for the unknown number of eggs: $(? \text{ eggs}) \cdot c = e \cdot 10c$. Divide both sides by c to get the solution: $(? \text{ eggs}) = \frac{10ce}{c} = 10e$.

13. Write the ratios as a proportion lining up the word labels to help. Convert 90 minutes to 1.5 hours so that the units used in both ratios are the same: $\frac{m \text{ miles}}{h \text{ hours}} = \frac{? \text{ miles}}{1.5 \text{ hours}}$. Cross multiply and solve for the unknown number of miles: $(? \text{ miles}) \cdot h = m \cdot 1.5$. Divide both sides

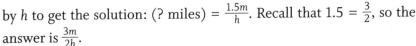

by h to get the solution: $(?\text{ miles}) = \frac{1.5m}{h}$. Recall that $1.5 = \frac{3}{2}$, so the answer is $\frac{3m}{2h}$.

14. Let x equal the width of the rectangle. Let $x + 3$ equal the length of the rectangle, because the length is "3 more than" the width. Because perimeter is the distance around the rectangle, the formula for perimeter (P) is length + width + length + width, or $P = l + w + l + w$, or $P = 2l + 2w$. Substitute the "let" statements for l and w and perimeter equal to 42 into the formula: $42 = 2(x + 3) + 2(x)$. Use the distributive property on the right side of the equation: $42 = 2x + 6 + 2x$. Combine like terms on the right side of the equation: $42 = 4x + 6$. Subtract 6 from both sides of the equation: $42 - 6 = 4x + 6 - 6$. Simplify: $36 = 4x$. Divide both sides of the equation by 4: $\frac{36}{4} = \frac{4x}{4}$. Therefore, $9 = x$.

15. Let w equal the width of the field and let $2w + 2$ equal the length of the field, because the length is two more than twice the width. Because the area of the field is known, use the formula $Area = length \times width$. Substitute to get the equation $1{,}200 = w(2w + 2)$. Use the distributive property on the right side of the equal sign: $1{,}200 = 2w^2 + 2w$. Subtract 1,200 from both sides to get the equation in standard form: $0 = 2w^2 + 2w - 1{,}200$. Factor the right side of the equation completely: $0 = 2(w^2 + w - 600)$, which becomes $0 = 2(w + 25)(w - 24)$. Set each factor equal to zero and solve: $2 \neq 0$; $w + 25 = 0$ or $w - 24 = 0$. The two values of w are -25 and 24. Because you can't have negative length, reject the value of -25. The width of the rectangular field is 24 feet.

16. Let w equal the width of the parallelogram and let l equal the length of the parallelogram. Because the length is 5 more than the width, then $l = w + 5$. The formula for the perimeter of a parallelogram $2l + 2w = 100$. Substituting the first equation into the second for l results in $2(w + 5) + 2w = 100$. Use the distributive property on the left side of the equation: $2w + 10 + 2w = 100$. Combine like terms on the left side of the equation: $4w + 10 = 100$. Subtract 10 from both sides of the equation: $4w + 10 - 10 = 100 - 10$. Divide both sides of the equation by 4: $\frac{4w}{4} = \frac{90}{4}$. So, $w = 22.5$. Therefore, the width is 22.5 centimeters and the length is $22.5 + 5 = 27.5$ centimeters.

17. Let x equal the number of hours they can remodel the kitchen if they work together. In one hour, Heather can do $\frac{1}{10}$ of the work and Steve can do $\frac{1}{15}$ of the work. As an equation, this information looks like $\frac{1x}{10} + \frac{1x}{15} = 1$, where

1 represents 100% of the work. Multiply both sides of the equation by the least common denominator, 30, to get the equation $3x + 2x = 30$. Combine like terms to get $5x = 30$. Divide both sides by 5: $\frac{5x}{5} = \frac{30}{5}$. The solution is 6 hours.

18. Let x equal the number of hours they need to seal the driveway if they work together. Because the time was given in minutes, divide by 60 to get the time in hours. In one hour, Anthony can do $\frac{1}{2.5}$ of the work and John can do $\frac{1}{2}$ of the work. As an equation, this information looks like $\frac{1x}{2.5} + \frac{1x}{2} = 1$, where 1 represents 100% of the work. Multiply both sides of the equation by the least common denominator, 10, to result in the equation $4x + 5x = 10$. Combine like terms to get $9x = 10$. Divide both sides by 9: $\frac{9x}{9} = \frac{10}{9}$. The solution is $1\frac{1}{9}$ hours.

Posttest

If you have completed all the chapters in this book, then you are ready to take the posttest to measure your progress. The posttest has 50 questions covering the topics you studied in this book. Although the format of the posttest is similar to that of the pretest, the questions are different.

Take as much time as you need to complete the posttest. When you are finished, check your answers with the answer key at the end of the posttest. Along with each answer is a number that tells you which chapter of this book teaches you the math skills needed for that question. Once you know your score on the posttest, compare the results with the pretest. If you scored better on the posttest than on the pretest, congratulations! You have profited from the hard work. At this point, you should look at the questions you missed, if any. Do you know why you missed the question, or do you need to go back to the chapter and review the concept?

If your score on the posttest doesn't show much improvement, take a second look at the questions you missed. Did you miss a question because of an error you made? If you can figure out why you missed the question, then you understand the concept and just need to concentrate more on accuracy when taking a test. If you missed a question because you did not know how to work a problem, go back to the chapter and spend time working that type of problem. Take time to understand algebra thoroughly. You need a solid foundation in algebra if you plan to use this information or progress to a higher level of algebra. Whatever your score on this posttest, keep this book for review and future reference.

1. Express three times the sum of a number (n) and 5.

2. If Carmen was x years old y years ago, how old will she be z years from now?

3. Simplify $a^2b \cdot ab^3$.

4. Simplify $\frac{x^6}{x^3}$.

5. Simplify $(3xy^3)^2$.

6. Which of the following is equivalent to $\sqrt{128}$?
   a. $12\sqrt{8}$
   b. $2\sqrt{8}$
   c. $8\sqrt{2}$
   d. $16\sqrt{8}$
   e. $64\sqrt{2}$

7. Which of the following is equivalent to $-3\sqrt{2} \cdot 7\sqrt{5}$?
   a. $10\sqrt{7}$
   b. $-10\sqrt{10}$
   c. $-21\sqrt{7}$
   d. $-21\sqrt{10}$
   e. $-\sqrt{210}$

8. How many solutions are there in a linear system of two distinct parallel lines?

9. How many solutions are there to a system of two distinct linear equations where the slopes of the lines are negative reciprocals of each other?

10. By solving the following system of equations by elimination method, what next step(s) would eliminate the variable y?
    $2x - 3y = 12$
    $-x + y = 4$

11. Solve the following system of equations by graphing.

$y = -x + 4$

$y = x$

12. Find the value of x in the solution to the system of equations algebraically by elimination.

$2x + y = 1$

$3x - 2y = 12$

13. Find the value of y in the solution to the system of equations algebraically by elimination.

$3x + 14 = 5y$

$2x + 7y = 1$

14. Find the value of x in the solution to the system of equations algebraically by substitution.

$y = 7x - 8$

$y + 4x = -19$

15. Find the value of y in the solution to the system of equations algebraically by substitution.

$-3x + y = 4$

$y = 4x + 1$

16. Find the solution to the following system of equations algebraically.

$3x + 4y = 6$

$2x - 6y = 4$

17. What is the solution to the following system of equations?

$-5y + x = 17$

$x = 2$

18. What is the solution set of the inequality $x - 3 > -8$?

19. What is the solution of the inequality $-\frac{a}{2} - 10 \geq -2$?

20. Solve the inequality: $x + 7 > 11$.

21. Solve the inequality: $9 - 5x \geq 39$.

22. Which ordered pair is in the solution set of the inequality $3x \geq -y - 6$?
   a. $(10,-10)$
   b. $(1,-10)$
   c. $(-3,-2)$
   d. $(-5,1)$
   e. $(0,-7)$

23. What is the degree of the monomial $10x^3y^2$?

24. Which of the following monomials has degree 3?
   a. $3x$
   b. $3x^2$
   c. $3xy$
   d. $3x^2y$
   e. $3x^3y$

25. What is the degree of the polynomial $5xy^2 + 3x$?

26. Simplify the expression $-4y - 3y + 2y$.

27. Simplify the expression $6a + 2 - 2a$.

28. Simplify the expression $2a^2b + 4a^2b + a^2b$.

29. Simplify the expression $3xy^2 + 4x^2y - xy^2$.

30. Which of the following shows the expression $7(x - 3) + 21$ in simplified form?
   a. $7x$
   b. $7x + 18$
   c. $10x + 21$
   d. $10x$
   e. $7x + 24$

31. Simplify the expression $(2a - b) + (7a + b)$.

32. Simplify the expression $9(2x - 1) - 2(x - 5)$.

33. What is the GCF of $3x^2$ and $9x^2y$?

34. Factor $12ab^2c - 16a^2bc - 4abc$ completely.

35. Factor the binomial $x^2 - 25$.

36. If $x - 1$ is one factor of the trinomial $3x^2 - 5x + 2$, what is the other factor?

37. Which of the following are factors of the binomial $x^2 + 16$?
   a. $(x + 4)(x + 4)$
   b. $(x - 4)(x - 4)$
   c. $(x + 8)(x + 2)$
   d. $(x - 8)(x - 2)$
   e. none of the above

38. Factor the binomial $2x^2 - 10xz$.

39. What are the roots of the equation $x^2 - 8x = -5$?

40. Peter has a rectangular yard whose length is 5 meters more than the width. If the area of his yard is 500 meters^2, what is the length of his yard?

41. The square of a number added to 64 equals 16 times the number. What is the number?

42. What is the greater of two consecutive positive odd integers whose sum is 256?

43. What is the lesser of two consecutive positive even integers whose product is 1,088?

44. Kala invested $2,000 in an account that earns 5% interest and x dollars in a different account that earns 7% interest. How much is invested at 7% if the total amount of interest earned is $310?

45. Sheila bought 10 CDs that cost d dollars each. What is the total cost of the CDs in terms of d?

46. Nancy and Jack can shovel the driveway together in 6 hours. If it takes Nancy 10 hours working alone, how long will it take Jack working alone?

47. What is $-5x(-2x^2 + 4x)$ equivalent to?

48. What is $(b + 3)(b + 7)$ equivalent to?

49. What is $(c + 6)(c^2 + c + 11)$ equivalent to?

50. Use the slope/y-intercept method to write an equation that would enable you to draw a graphic solution to the following word problem: An Internet service provider charges $15 plus $0.25 per hour of usage per month.

ANSWERS

1. Three times the sum of a number (n) and 5 is $3n + 5$. For more help with this concept, see Chapter 3.

2. If Carmen was x years old y years ago, then she is $x + y$ years old now, and she will be $x + y + z$ years old z years from now. For more help with this concept, see Chapter 3.

3. When multiplying like bases, add the exponents. The expression $a^2b \cdot ab^3$ can also be written as $a^2b^1 \cdot a^1b^3$. Grouping like bases results in $a^2a^1 \cdot b^1b^3$. Adding the exponents gives $a^{(2+1)}b^{(1+3)}$, which is equal to a^3b^4, the simplified answer. For more help with this concept, see Chapter 5.

4. When dividing like bases, subtract the exponents. The expression $\frac{x^6}{x^3}$ then becomes $x^{(6-3)}$, which simplifies to x^3. For more help with this concept, see Chapter 5.

5. When raising a quantity to a power, raise each base to that power by multiplying the exponents. The expression $(3xy^3)^2$ equals $3^2x^2y^6$, which simplifies to $9x^2y^6$. Another way to look at this problem is to remember that when a quantity is squared, it is multiplied by itself. The expression $(3xy^3)^2$ becomes $(3xy^3) \cdot (3xy^3)$. Multiply coefficients and add the exponents of like bases. $3 \cdot 3x^{(1+1)}y^{(3+3)}$ simplifies to $9x^2y^6$. For more help with this concept, see Chapter 5.

6. To simplify a radical, find the largest perfect square factor of the radicand. Because 128 can be expressed as $64 \cdot 2$, then write $\sqrt{128}$ as $\sqrt{64} \cdot \sqrt{2}$. Because $\sqrt{64}$ equals 8, the radical reduces to $8\sqrt{2}$. For more help with this concept, see Chapter 5.

7. When multiplying radicals, multiply the numbers in front of the radicals together and then the radicands together. For the expression $-3\sqrt{2} \cdot 7\sqrt{5}$, multiply $-3 \cdot 7$ and $\sqrt{2} \cdot \sqrt{5}$. This simplifies to $-21\sqrt{10}$. For more help with this concept, see Chapter 5.

8. There is no solution because the lines will never cross. Two lines that are parallel have the same slope and will never intersect. For more help with this concept, see Chapter 6.

9. This type of system will have one solution. Because the slopes are negative reciprocals of each other, they will meet to form right angles in one location. This point of intersection is the solution to the system. For more help with this concept, see Chapter 7.

10. Eliminate y by getting coefficients of y that are inverses, like -3 and 3. Multiplying the second equation by 3 will change it to $-3x + 3y = 12$. Adding the two equations will now eliminate the y terms: $-3y + 3y = 0y = 0$. For more help with this concept, see Chapter 4.

11. In the equation $y = -x + 4$, the slope is -1 and the y-intercept is 4. In the equation $y = x$, the slope is 1 and the y-intercept is 0. Your graph should show the graph of the equations on the same set of axes. The two lines intersect at the point $(2,2)$. This is the solution to the system of equations. For more help with this concept, see Chapter 6.

12. Because you are looking for the value of x, eliminate the variable y. First, multiply the first equation by 2 to get the coefficients of y to be opposites. Then, add the two equations vertically.
$2(2x + y = 1) \rightarrow 4x + 2y = 2$
$3x - 2y = 12 \rightarrow \underline{3x - 2y = 12}$
$7x = 14$ (Recall that $2y + -2y = 0$.)
Because $7x = 14$, divide each side by 7 to get $x = 2$.
For more help with this concept, see Chapter 7.

13. Because you are looking for the value of y, eliminate the variable x. First, rewrite the first equation as $3x - 5y = -14$ to line up like terms. Multiply the first equation by 2 and the second equation by -3 to get the coefficients of x to be opposites. Then, add the two equations vertically.
$2(3x - 5y = -14) \rightarrow 6x - 10y = -28$
$-3(2x + 7y = 1) \rightarrow \underline{-6x - 21y = \quad -3}$
$-31y = -31$
Because $-31y = -31$, divide each side by -31 to get $y = 1$.
For more help with this concept, see Chapter 7.

14. Because the variable y is isolated in the first equation, substitute $7x - 8$ for y in the second equation. The equation $y + 4x = -19$ becomes $(7x - 8) + 4x = -19$. Combine like terms on the left side of the equation: $11x - 8 = -19$. Add 8 to both sides: $11x - 8 + 8 = -19 + 8$. This simplifies to $11x = -11$. Divide both sides by 11: $\frac{11x}{11} = \frac{-11}{11}$; $x = -1$. For more help with this concept, see Chapter 7.

15. Because y is isolated in the second equation, substitute $4x + 1$ for y in the first equation. The equation $-3x + y = 4$ becomes $-3x + (4x + 1) = 4$. Combine like terms on the left side of the equation: $x + 1 = 4$. Subtract 1 from both sides of the equation. Therefore, $x = 3$. To solve

for y, substitute $x = 3$ into $y = 4x + 1$. So, $y = 4(3) + 1 = 12 + 1 = 13$. For more help with this concept, see Chapter 7.

16. Because there are many different coefficients in both equations, use the elimination method. Multiply the first equation by 3 and the second equation by 2 to get the coefficients of y to be opposites. Then, add the two equations vertically.

    $3(3x + 4y = 6) \rightarrow 9x + 12y = 18$
    $2(2x - 6y = 4) \rightarrow \underline{4x - 12y = 8}$
    $13x = 26$

    Because $13x = 26$, divide each side by 13 to get $x = 2$. This question asks for the solution to the system, so substitute $x = 2$ into the first equation. The equation $3x + 4y = 6$ becomes $3(2) + 4y = 6$. This simplifies to $6 + 4y = 6$. Subtract 6 from both sides of the equation: $4y = 0$. So, $y = 0$, and the solution is $(2,0)$. For more help with this concept, see Chapter 7.

17. Because $x = 2$ is one of the equations, substitute 2 for x in the first equation: $-5y + x = 17$ becomes $-5y + 2 = 17$. Subtract 2 from both sides of the equation: $-5y + 2 - 2 = 17 - 2$. This simplifies to $-5y = 15$. Divide both sides by -5: $\frac{-5y}{-5} = \frac{15}{-5}$. So, $y = -3$. Because $x = 2$ and $y = -3$, the solution to the system is $(2,-3)$. For more help with this concept, see Chapter 7.

18. Solve the inequality $x - 3 > -8$ as you would an equation. Add 3 to both sides of the inequality: $x - 3 + 3 > -8 + 3$. This simplifies to $x > -5$. For more help with this concept, see Chapter 8.

19. Solve the inequality as you would an equation. Add 10 to both sides of the inequality: $-\frac{a}{2} - 10 + 10 \geq -2 + 10$. Simplify: $-\frac{a}{2} \geq 8$. Multiply each side by -2: $-\frac{a}{2} \cdot -2 \leq 8 \cdot -2$. Notice the reversed inequality symbol and simplify: $a \leq -16$. For more help with this concept, see Chapter 8.

20. Solve the inequality for x. Subtract 7 from each side of the inequality: $x + 7 - 7 > 11 - 7$. This simplifies to $x > 4$. For more help with this concept, see Chapter 8.

21. Solve the inequality for x. Subtract 9 from both sides of the inequality: $9 - 9 - 5x \geq 39 - 9$, or $-5x \geq 30$. Now divide both sides by -5. Don't forget to flip the inequality sign because you are dividing by a negative number. $\frac{-5x}{-5} \leq \frac{30}{-5} = x \leq -6$. For more help with this concept, see Chapter 8.

22. One way to solve this problem is to substitute each ordered pair into $3x \geq -y - 6$ and find the true inequality. With $(10, -10)$, the inequality becomes $3(10) \geq -(-10) - 6$. This simplifies to $30 \geq 10 - 6$, which is equal to $30 \geq 4$. This is true, so $(10, -10)$ is the correct answer. Another way to solve this problem is to graph the original inequality and the points from each of the answer choices. Only the point $(10, -10)$ would be located in the shaded region of the inequality. For more help with this concept, see Chapter 8.

23. The degree of a monomial is the sum of the exponents on the variables. The exponent on x is 3, and the exponent on y is 2. So, $3 + 2 = 5$; therefore, the degree is 5. For more help with this concept, see Chapter 9.

24. The degree is the sum of the exponents on the variables; $3x$ has degree 1. $3x^2$ has degree 2; $3xy$ has degree 2; $3x^2y$ has degree 3, so this is the correct answer. The exponents are 2 and 1. So, $2 + 1 = 3$. The degree of $3x^3y$ is 4. For more help with this concept, see Chapter 9.

25. The degree of a polynomial is the degree of the term with the greatest degree. The first term in the polynomial has degree 3 because the exponents are 1 and 2. The second term has degree 1. The term with the greatest degree has a degree of 3, so the polynomial has degree 3. For more help with this concept, see Chapter 9.

26. Add and subtract like terms in order from left to right: $-4y - 3y + 2y$. Change subtraction to addition and reverse the sign of the following term: $-4y + -3y + 2y$. The signs are different, so subtract the coefficients: $-7y + 2y = -5y$. For more help with this concept, see Chapter 3.

27. Use the commutative property to arrange like terms together: $6a - 2a + 2$. Subtract like terms: $6a - 2a + 2 = 4a + 2$. For more help with this concept, see Chapter 3.

28. All of the terms are like terms because the variables in each are exactly the same, including the exponents. To simplify the expression, combine the coefficients: $2 + 4 + 1 = 7$. Keep the variables that were originally with each term to make $7a^2b$. For more help with this concept, see Chapter 3.

29. Because not every term has exactly the same variable and exponent configuration, you cannot just combine the coefficients here. You may combine like terms only, which are the first and third terms: $3xy^2$ and $-xy^2$. Use the commutative property to put together like terms: $3xy^2 - xy^2 + 4x^2y$. Subtract like terms: $2xy^2 + 4x^2y$. For more help with this concept, see Chapter 3.

30. Eliminate the parentheses first by using the distributive property. The expression $7(x - 3) + 21$ becomes $7x - 21 + 21$. Combine like terms: $7x + 0 = 7x$. For more help with this concept, see Chapter 3.

31. Use the commutative property of addition to group together like terms: $2a + 7a - b + b$. Combine like terms: $9a$. Notice that $-b + b = 0b = 0$. For more help with this concept, see Chapter 3.

32. Change subtraction signs to addition and reverse the sign of the number that follows: $9(2x + -1) + -2(x + -5)$. Eliminate both sets of parentheses first by using the distributive property: $18x + -9 + -2x + 10$. Use the commutative property of addition to group together like terms: $(18x + -2x) + (-9 + 10)$. Combine like terms: $16x + 1$. For more help with this concept, see Chapter 3.

33. The GCF is the greatest common factor of both terms. Each term has a numerical factor of 3 and a variable factor of x^2. Therefore, the GCF is $3x^2$. For more help with this concept, see Chapter 9.

34. In order to factor this trinomial completely, first look for the greatest common factor, or GCF, of the three terms. Each term contains a numerical factor of 4 and variable factors of abc. Therefore, the GCF is $4abc$. If you divide each term by $4abc$, you are left with $3b - 4a - 1$ in the parentheses. Therefore, the factored form is $4abc(3b - 4a - 1)$. For more help with this concept, see Chapter 9.

35. This binomial is the difference between two perfect squares. The square root of x^2 is x and the square root of 25 is 5. The factors expressed as the sum and the difference of these square roots are $(x + 5)$ and $(x - 5)$. For more help with this concept, see Chapter 9.

36. In this case, you know one of the factors of the trinomial. To find the other factor, first use the fact that the first term in the trinomial is $3x^2$. (x times $3x$ equals $3x^2$.) The product of the last terms in each of the binomials is equal to 2 and the known factor is $x - 1$, so -1 would be multiplied by -2 to get a value of $+2$. This leads you to $3x - 2$ as the solution. Use FOIL (distributive property) to check your answer: $(x - 1)(3x - 2) = 3x^2 - 2x - 3x + 2 = 3x^2 - 5x + 2$. For more help with this concept, see Chapter 9.

37. This binomial is not factorable. It is the *sum* of two perfect squares. Do not confuse it with $x^2 - 16$, which is factorable because it is the *difference* between two perfect squares. If you multiply out any of the choices,

none of them will equal $x^2 + 16$. For more help with this concept, see Chapter 9.

38. First, look for factors that both terms have in common. The terms $2x^2$ and $10xz$ both have a factor of 2 and x. Factor out the greatest common factor, $2x$, from each term: $2x^2 - 10xz = 2x(x - 5z)$. To check an answer like this, multiply through using the distributive property: $2x(x - 5z) = 2x \cdot x - 2x \cdot 5z = 2xz - 10xz$. This question checked because the result is the same as the original binomial. For more help with this concept, see Chapter 9.

39. Put the equation in standard form: $x^2 - 8x + 5 = 0$. Because this equation is not factorable, use the quadratic formula by identifying the value of a, b, and c and then by substituting them into the formula. For this particular equation, $a = 1$, $b = -8$, and $c = 5$. The quadratic equation is $x = \frac{-b \pm \sqrt{b^2 - 4ac}}{2a}$. Substitute the values of a, b, and c to get $x = \frac{-(-8) \pm \sqrt{(-8)^2 - 4(1)(5)}}{2(1)}$. This simplifies to $x = \frac{8 \pm \sqrt{64 - 20}}{2}$, which becomes $x = \frac{8 \pm \sqrt{44}}{2}$. So, $x = \frac{8}{2} \pm \frac{2\sqrt{11}}{2}$. After you reduce this result, the solution is $x = \{4 - \sqrt{11}, 4 + \sqrt{11}\}$. For more help with this concept, see Chapter 9.

40. Let x equal the width of his yard. Therefore, $x + 5$ equals the length of the yard. Because the area is 500 meters^2, and area is length times width, the equation is $x(x + 5) = 500$. Use the distributive property to multiply the left side: $x^2 + 5x = 500$. Subtract 500 from both sides: $x^2 + 5x - 500 = 500 - 500$. Simplify: $x^2 + 5x - 500 = 0$. Factor the result: $(x - 20)(x + 25) = 0$. Set each factor equal to zero and solve: $x - 20 = 0$ or $x + 25 = 0$. The solution is $x = 20$ or $x = -25$. Reject the solution of -25 because a distance can never be negative. Because the width is 20 meters, the length is 25 meters. For more help with this concept, see Chapter 10.

41. Let x equal the number. The statement "The square of a number added to 64 equals 16 times the number" translates into the equation $x^2 + 64 = 16x$. Put the equation in standard form and set it equal to zero: $x^2 - 16x + 64 = 0$. Factor the left side of the equation: $(x - 8)(x - 8) = 0$. Set each factor equal to zero and solve: $x - 8 = 0$ or $x - 8 = 0$. The solution is $x = 8$ or $x = 8$, so the number is 8. For more help with this concept, see Chapter 10.

42. Let x equal the lesser odd integer, and let $x + 2$ equal the greater odd integer. Because *sum* is a key word for addition, the equation is $x + x + 2 = 256$. Combine like terms on the left side of the equation: $2x + 2 = 256$. Subtract 2 from both sides of the equation: $2x + 2 - 2 = 256 - 2$. This simplifies to $2x = 254$. Divide each side by 2: $\frac{2x}{2} = \frac{254}{2}$. So, $x = 127$. Thus, the greater odd integer is $x + 2 = 129$. For more help with this concept, see Chapter 10.

43. Let x equal the lesser even integer, and let $x + 2$ equal the greater even integer. Because *product* is a key word for multiplication, the equation is $x(x + 2) = 1{,}088$. Multiply using the distributive property on the left side of the equation: $x^2 + 2x = 1{,}088$. Put the equation in standard form and set it equal to zero: $x^2 + 2x - 1{,}088 = 0$. Factor the trinomial: $(x - 32)(x + 34) = 0$. Set each factor equal to zero and solve: $x - 32 = 0$ or $x + 34 = 0$. So, $x = 32$ or $x = -34$. Because you are looking for a positive integer, reject the x-value of -34. Therefore, the lesser positive integer would be 32. For more help with this concept, see Chapter 10.

44. Let x equal the amount invested at 7% interest. Because the total interest is $310, use the equation $0.05(2{,}000) + 0.07x = 310$. Multiply: $100 + 0.07x = 310$. Subtract 100 from both sides: $100 - 100 + 0.07x = 310 - 100$. Simplify: $0.07x = 210$. Divide both sides by 0.07: $\frac{0.07x}{0.07} = \frac{210}{0.07}$. Therefore, $x = \$3{,}000$, which is the amount invested at 7% interest. For more help with this concept, see Chapter 10.

45. Suppose that the cost for one CD, or d, is $15. Then the total cost of 10 CDs is $15 \cdot 10$, by multiplying the cost per CD by how many CDs are purchased. Therefore, in terms of d, the total cost is $d \cdot 10$, which is equivalent to $10d$. Another way to look at this problem is to write the ratios as a proportion lining up the word labels to help: $\frac{1\,CD}{d\,dollars} = \frac{10\,CDs}{?\,dollars}$. Cross multiply and solve for the unknown number of dollars: $(?\,dollars) \cdot 1 = 10 \cdot d$. Therefore, the total cost is $10d$. For more help with this concept, see Chapter 10.

46. Let x equal the number of hours Jack takes to shovel the driveway by himself. In 1 hour, Jack can do $\frac{1}{x}$ of the work and Nancy can do $\frac{1}{10}$ of the work. As an equation, this information looks like $\frac{1}{x} + \frac{1}{10} = \frac{1}{6}$, where $\frac{1}{6}$ represents what part they can shovel together in one hour. Multiply both sides of the equation by the least common denominator, $30x$, to get an equation of $30 + 3x = 5x$. Subtract $3x$ from both sides of the equation: $30 + 3x - 3x = 5x - 3x$. This simplifies to $30 = 2x$: Divide both sides of the equal sign by 2 to get a solution of 15 hours. For more help with this concept, see Chapter 10.

47. Distribute the $-5x$ inside the parentheses: $-5x(-2x^2 + 4x)$ becomes $10x^3 - 20x^2$. For more help with this concept, see Chapter 9.

48. Use FOIL (First, Outer, Inner, and Last): $(b + 3)(b + 7)$ becomes $b^2 + 7b + 3b + 21 = b^2 + 10b + 21$. For more help with this concept, see Chapter 9.

49. Both of the terms of the binomial need to be distributed inside the parentheses of the trinomial: $(c + 6)(c^2 + c + 11)$ becomes $c(c^2) + c(c) + 11(c) + 6(c^2) + 6(c) + 6(11) = c^3 + c^2 + 11c + 6c^2 + 6c + 66$. Combine like terms: $c^3 + 7c^2 + 17c + 66$. For more help with this concept, see Chapter 9.

50. Let y equal the amount of a monthly bill. Let x equal the hours of Internet use for the month. The costs for the month will equal \$15 plus the \$0.25 times the number of hours of use. Written as an equation, this information would be as follows: $y = 0.25x + 15$. A graph of this equation would have a slope of 0.25, or $\frac{25}{100} = \frac{1}{4}$. The y-intercept would be at $(0,15)$. For more help with this concept, see Chapter 6.

Glossary

Additive property of zero: The rule that when you add zero to a number, the result is that number. For example, $-6 + 0 = -6$ or $x + 0 = x$.

Binomial: An expression with two terms. For example, $a + b$.

Coefficient: The number in front of the variable(s).

Commutative property: The rule that allows you to change the order of the numbers when you add or multiply.

Coordinate plane: A graph formed by two lines that intersect to create right angles.

Distributive property: The rule that allows you to multiply the number and/or variable(s) outside the parentheses by every term inside the parentheses. For example, $2(a - b + 3) = 2a - 2b + 6$; $3x(x + 2) = 3x^2 + 6x$.

Equation: Two equal expressions. Examples: $2 + 2 = 1 + 3$ and $2x = 4$.

Evaluate: To substitute a number for each variable and simplify.

Exponent: A number that tells you how many times a factor is multiplied. An exponent appears smaller and raised. Example: $2^3 = 2 \cdot 2 \cdot 2$.

Factors: Numbers to be multiplied in a multiplication operation. Example: The numbers 2 and 6, when multiplied together, are factors of 12.

Inequality: Two expressions that are not equal and are described by an inequality symbol such as $<$, $>$, $\leq$, $\geq$, or $\neq$.

Integers: All the whole numbers and their opposites. Integers do not include fractions. The integers can be represented in this way: . . . $-3, -2, -1, 0, 1, 2, 3$ . . .

Like terms: Terms that have the same variable(s) with the same exponent. Example: $3x^2y$ and $5x^2y$.

Linear equation: A linear equation always graphs into a straight line. The variable in a linear equation cannot contain an exponent greater than one. This type of equation cannot have a variable in the denominator, and the variables cannot be multiplied.

Linear inequality: An inequality that has one of the four forms below:

$ax + by < c$

$ax + by > c$

$ax + by \leq c$

$ax + by \geq c$

The polynomials in a linear inequality cannot have exponents greater than one.

Monomial: An expression with one term.

Order of operations: The sequence of performing steps to get the correct answer. The order you follow is

1. Simplify all operations within grouping symbols such as parentheses, brackets, braces, and fraction bars.
2. Evaluate all exponents.
3. Do all multiplication and division in order from left to right.
4. Do all addition and subtraction in order from left to right.

Ordered pair: Two numbers in a specific sequence that represent a point on a coordinate plane. The numbers are enclosed in parentheses with the x-coordinate first and the y-coordinate second; for example, (2,3).

Origin: The starting point, or zero, on a number line. On a coordinate plane, the origin is the point where the x- and y-axes intersect. The coordinates of the origin are (0,0).

Polynomial: A polynomial is a number, variable, or combination of a number and a variable. Examples: 5, $3x$, and $2x + 2$.

Quadrants: The four equal parts of a coordinate plane. A number names each quadrant. The upper-right-hand area is quadrant I. You proceed counterclockwise to name the other quadrants.

Quadratic equation: An equation in which the highest power of the variable is 2. The graph of a quadratic equation is a smooth curve. A quadratic equation is represented by $ax^2 + bx + c = 0$.

Radical equation: An equation that has a variable in the radicand.

Radical sign: The mathematical symbol that tells you to take the root of a number.

Radicand: The number under the radical sign in a radical.

Slope: The steepness of a line. Slope is the rise over the run or the change in y over the change in x.

Slope-intercept form: $y = mx + b$. Also known as $y =$ form.

Square root: The opposite of a square of a number. The square root of 16 is 4 because 4 times 4 equals 16; expressed as $\sqrt{16} = 4$.

Squaring a number: Multiplying a number by itself. Example: 4×4.

System of equations: Two or more equations with the same variables.

System of inequalities: Two or more inequalities with the same variables.

Trinomial: An expression with three terms. For example, $a + b + c$.

Variable: A letter representing a number.

Whole numbers: 0, 1, 2, 3, . . . Whole numbers start with 0 and do not include fractions.

x-axis: The horizontal line that passes through the origin on the coordinate plane.

y-axis: The vertical line that passes through the origin on the coordinate plane.

y-intercept: The point where a line graphed on the coordinate plane intersects the y-axis.

Zero product property: The rule that when the product of two numbers is zero, then one or both of the factors must equal zero. For example, $ab = 0$ if $a = 0$, $b = 0$, or both equal 0.

Notes

Notes

Notes

Notes

Notes

Notes